Thanks for Everything (Now Get Out)

THANKS FOR EVERYTHING (NOW GET OUT)

CAN WE RESTORE NEIGHBORHOODS WITHOUT DESTROYING THEM?

JOSEPH MARGULIES

Yale

UNIVERSITY PRESS

New Haven and London

Published with assistance from the foundation established in
memory of Calvin Chapin of the Class of 1788, Yale College.

Yale University Press books may be purchased in quantity for
educational, business, or promotional use. For information, please e-mail
sales.press@yale.edu (U.S. office) or sales@yaleup.co.uk (U.K. office).

Set in Adobe Garamond and Gotham types
by Tseng Information Systems, Inc.
Printed in the United States of America.

Library of Congress Control Number: 2021935429
ISBN 978-0-300-25001-5 (hardcover : alk. paper)

A catalogue record for this book is available from the British Library.

This paper meets the requirements of ANSI/NISO z39.48-1992
(Permanence of Paper).

10 9 8 7 6 5 4 3 2 1

To Sandra, with all my love

Contents

Olneyville

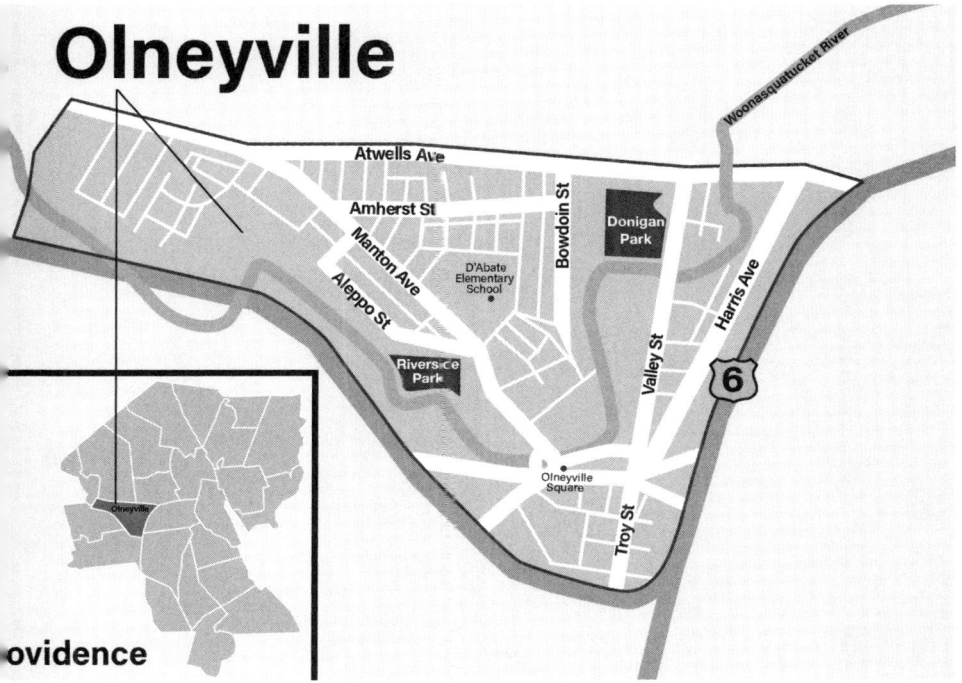

Olneyville, a neighborhood in Providence, Rhode Island.
Map created by Zoe Ambinder.

Prologue

Imagine you want to gentrify a low-income urban neighborhood and displace its residents. I realize that's probably not your ambition. In fact, I suspect you're reading this book because, like me, you want to prevent gentrification and displacement. But for a minute, just imagine that's your goal. How would you go about it?

The first thing you'd do is find the right neighborhood. Ideally, you'd pick somewhere close to downtown, with major transportation systems already in place so that people could get around easily, preferably without a car. It would also be nice if the neighborhood were built around a natural attraction, like a river or a rise in the land, or had a distinctive history that gave it a unique cultural identity. And, of course, you'd want someplace that was run-down, where land was cheap. This would allow you to maximize your profit when things turned around.

Once you found the right location, you'd want someone to fix it up. The empty factories, abandoned houses, and boarded-up stores—they would have to be torn down or rebuilt. That toxic waste dump where an old plant burned down—that would have to be cleaned up. The violent crime and drug dealing—you'd definitely want somebody to get that under control. The police department might need a complete makeover. No more brutality or corruption. And you'd want someone to fix the local school, creating a place where parents would be happy to send their children. In short, you'd want an exciting, vibrant neighborhood with low crime rates, safe streets, affordable housing, respectful police, and a good public school. Is that so hard?

Now imagine this run-down neighborhood was your home. Imagine that you had lived there for years, and you and your neighbors were tired of the vacant lots, empty houses, and abandoned factories. You were sick of watching your kids make a playground out of a toxic dump, and fed up with the crime. You'd had enough of the police brutality and official corruption. You and your allies wanted to repair the neighborhood and create an exciting, healthy place for you and your children with safe streets, affordable housing, respectful police, and a good public school. Is that so hard?

This simple thought experiment reveals the dilemma at the core of this book. In a back-to-the-city moment, is gentrification the inevitable cost imposed on the poor for making their neighborhoods safe and vibrant? Can a distressed urban neighborhood be transformed without being destroyed?

This book is about Olneyville, a low-income, predominately Latino neighborhood on the west side of Providence, Rhode Island. It is based on interviews and research I conducted there from the summer of 2017 to early 2020. Not long ago, Olneyville was one of the most distressed neighborhoods in Rhode Island. Today it is on the cusp of gentrification. Junkyards have become parks, vacant lots have become affordable houses, crime is down, and the local public school is one of the most beloved in the city. And as the neighborhood changed, outsiders began to take notice. In the last decade, Olneyville has become wealthier, whiter, and more unequal.

What has happened (and is happening) in Olneyville is also happening in low-income minority neighborhoods in cities across the country: they are getting better, and the improvement is paving the way for gentrification and displacement. This book is about them too. It is about the conditions in these neighborhoods, from public space and housing to policing and schools, and the efforts that residents and their allies make to change them. It is also about the unintended effects of those changes. Success summons failure, as gentrification displaces the people who suffered through the neighborhood at its worst and worked so hard to make it better. And finally, it is about what we can do to prevent this from happening.

I write about neighborhoods rather than communities because people belong to many communities, but most of us live in only one place. In cities, we call that place a neighborhood. When we ask whether the place that people call home can be transformed but not destroyed, we are asking about their neighborhoods. This is particularly true in Providence, which is divided into twenty-five neighborhoods that define how residents describe where they live. People tend to say they live or grew up in, for instance, Olneyville, Mount Pleasant, or Smith Hill. Communities, by contrast, embrace much more than a place. In no particular order, I am a husband, a father, a lawyer, a professor, a Jew, an Italophile, a long-distance cyclist, and doubtless a few others that escape me right now, all of which provide me a connection to different communities. I am also a U.S. citizen who lives in Ithaca, New York. Some of my communities provide a connection to a place, but most do not, and what is true for me is true for most people.[1]

But why Olneyville? I was asked again and again during my research how I came to write about a small neighborhood of roughly seven thousand people in a midsize city in the smallest state in the country. Yet Olneyville was the perfect choice for a book that asks whether neighborhood transformation must give way to destruction. Its compact size—about half a square mile—made it easy to navigate by car or on foot. I could, and often did, crisscross the neighborhood many times in a single afternoon. What's more, the neighborhood's boundaries are almost identical to the boundaries of the census tract in which it is located, so that data available from the Census Bureau are a close proxy for the neighborhood itself. This allowed me to track changes in the neighborhood over time, which is especially important when we ask about encroaching gentrification and displacement. In addition, the current approach to neighborhood well-being nation-wide has had enviable success in Olneyville. As a result, the neighborhood shows us not just the potential of the model but its limits as well. If the current approach cannot save a distressed neighborhood even when it works as well as can be hoped, what chance does it have as a nationwide strategy?

Finally and most importantly, Olneyville makes a perfect case study because the damage is not yet done. In many places, gentrifica-tion and displacement have advanced so far that talking about saving the low-income neighborhoods that once existed there is pointless. They have been destroyed. But these forces are still in their early stages in Olneyville. We still have time to make a difference. In this, Olneyville is again representative. Skyrocketing costs in New York, Boston, San Francisco, and Washington, D.C., have shifted the de-bate about gentrification and displacement into smaller cities that were once considered immune to the pressures of the back-to-the-city movement. Neighborhoods in places like Nashville, Durham (N.C.), and Denver are the new battlegrounds.

Unless we change course, that battle will be lost. Neighborhoods like Olneyville *can* transform themselves. The people who live there do not have to settle for boarded-up houses and broken lives, as if de-spair were a law of nature in a low-income neighborhood. Nor does positive change have to pave the way for low-income residents to be pushed out. Neighborhoods in this country can be safe, viable, and affordable. But the transformation will not happen unless distressed neighborhoods wield tools that empower residents and bind capital. The present approach to neighborhood well-being will not supply those tools. We have to create them.

1

"Today Was a Good Day"

When Yuselly Mendoza was in college, she began each weekday morning with a walk through Olneyville. At 8:05, she and another woman donned bright-yellow jackets and set out from the corner of Salmon Street and Manton Avenue, across the street from Sanchez Liquors and a vacant, weed-choked lot. They followed a route that meandered through the neighborhood, stopping along the way to pick up schoolchildren who walked with them before finally reaching William D'Abate Elementary School. Around thirty-five kids accompanied them on the Walking School Bus, joining either with Yuselly and her companion on their walk ("the blue route") or with two other adults who followed a different path ("the green route"). Every afternoon at 3:30, Yuselly and the other women retraced their steps as they walked the children home. More kids would have walked with the bus if not for rules that required a certain ratio of adults to children. Since the school lacked the funding to hire another adult, the bus had a waiting list.

As simple as it is, the Walking School Bus opens a window on the enormous complexity of neighborhood well-being. The bus is about education, of course. It's about getting kids to their elementary school on time, which matters as much to parents in Olneyville as it does to parents anywhere. Though Olneyville is one of the poorest neighborhoods in Providence, D'Abate Elementary consistently has one of the highest attendance rates in the city. The bus is also about public safety.

Like many poor urban neighborhoods, Olneyville has more than its share of challenges, and the group walks past empty lots, abandoned buildings, and derelict houses. "The police used to be at this house all the time," Yuselly told me as we passed a vacant property with the windows boarded shut and the electrical boxes stripped bare.

The bus is also about public health. The green and blue routes are each about a mile long, and since the kids walk to and from school, they get some exercise and time outdoors. And in a less obvious way, the bus is about work, since the Walking School Bus gives parents the freedom and peace of mind to leave for their jobs without wondering whether their kids will get to school safely and on time. Yuselly told me she changed the start time for her route from 8:20 to 8:05 to accommodate one of the parents on the line, who had to be at work at 8:30 and wanted to be home when her child left the house.

Ultimately, the Walking School Bus is about the idea of community itself. It's a creative and visible demonstration of communal self-help and intricate relationship building—about neighbors providing transportation without vehicles, public safety without police, and public health without doctors or nurses. Yuselly, the bus "captain" from its earliest days in 2016, got to know the children on her route, and when one of them missed a couple of days, she made a point of alerting the child's teacher. Before long, someone had called the home or paid a visit to make sure everything was okay.

The Walking School Bus also reminds us that neighborhood well-being never comes down to just one characteristic. The many people and institutions that combine to create a thriving neighborhood are interwoven and mutually dependent, and a small, inexpensive, local initiative can have an outsize effect by rippling and compounding through the neighborhood. Imagine what would happen to the fabric of the neighborhood—to public health, public safety, education, jobs—if the Walking School Bus stopped running. Now multiply that by a hundred small programs that lace through a low-income neighborhood. When it comes to neighborhood well-being, everything is connected, and support for—or neglect of—one program affects all the others.

For decades, this interconnectedness has confounded attempts to transform distressed neighborhoods. Olneyville's Walking School Bus captures the current approach to this puzzle. The bus is funded by the Rhode Island Department of Health (RIDOH), which received a grant from the Centers for Disease Control and Prevention. RIDOH, in turn, funds nine organizations across the state, including ONE Neighborhood Builders, a community development corporation with a long record of collaborative community building in Olneyville. ONE Neighborhood acts as a backbone organization for about a dozen other nonprofits. It disburses funds, provides organizational support, and takes the laboring oar in building relationships among various groups, of which the bus is one. The entire model uses a nonconfrontational approach that distributes meager pools of money from foundations, philanthropies, and anchor institutions like universities and hospitals, as well as extremely limited public-sector funding, to an eclectic array of chronically underfunded nonprofit organizations that act on behalf of distressed neighborhoods.

Over the years, people have given this approach to neighborhood well-being a variety of names. Bruce Katz and Jeremy Nowak of the Brookings Institution, for instance, call it the New Localism.[1] I am not satisfied with this label, since the resources that flow into a distressed neighborhood are not at all locally controlled. Certainly the men and women who live in Olneyville would scoff at any suggestion that they manage or own the resources deployed on their behalf. Other writers have described the current approach as "the nonprofit industrial complex."[2] But the critique implicit in this name is aimed beyond neighborhood well-being to the entire world of nonprofits.

If we have to give the current model a name, I would call it *neoliberal precarity,* which alerts us to the urgency of the crisis in distressed neighborhoods and points us toward the primary culprit. Still, it is less important to name the problem than to solve it, and to do that, it helps to know where it came from.

Neighborhoods like Olneyville do not just *happen.* They are the predictable result of specific public and private choices. In broad terms, the history of Olneyville is the history of thousands of urban neighborhoods across the country: postwar suburbanization encouraged by the national housing policy; urban job loss and disinvestment; social and economic dislocation caused by years of highway construction and "urban renewal"; collapsing infrastructure; and rising levels of crime and disorder. It is the story of low-income urban America in the second half of the twentieth century.[3] Because of the city's overreliance on the textile industry, Providence collapsed sooner and harder than many other industrial cities. Olneyville followed the same fate, becoming increasingly segregated and socially isolated. Over the past three decades, its racial and ethnic mix has shifted from white to Latino, and its isolation has intensified as the neighborhood became home to more residents who are undocumented and speak English only as a second (or third) language.

As neighborhoods like Olneyville collapsed, the ideology of neoliberalism rose. At its simplest, neoliberalism is the idea that social problems are better solved by the private sector than by government. But neoliberalism is more than an economic or political model. It is a mind-set, a way to make sense of the world. Government, so the neoliberals argue, cannot solve social problems because it is too inefficient, bloated, and sclerotic. But more importantly, it *should not.* Government support creates dependence and stifles personal drive and ambition. People are truly free only when they take responsibility for their own lives. If a single mother lives in substandard housing or an unsafe neighborhood, if she sends her children to failing schools or is stuck in a low-wage job, it is because she made bad choices — about reproduction, education, housing, her profession. Because she did not take responsibility for herself and her family, she must endure the consequences. At least that is the neoliberal argument.[4]

Neoliberalism conditions us to train our attention on individual actors and their "good" or "bad" choices rather than on structural conditions that constrain or eliminate choice altogether. Worse, by reducing urban life to a scramble for personal gain — securing a spot in the top school or an apartment on the best block — neoliberalism drives people to fend for themselves, leaving the most vulnerable unprotected and destroying the very idea of shared obligation.

In a neoliberal age, the principal role of government in a distressed neighborhood like Olneyville is to create opportunities for private investors. Take housing, for instance. The federal government has not built public housing for decades. Instead it uses tax breaks or subsidies to create financial incentives for landlords and developers to provide shelter for some (but nowhere near all) of the people who cannot afford housing at market rates. Responsibility for housing these people has shifted from government itself to private actors who are incentivized by government to create the appearance of "choice" for low-income residents.

Many people refer to neoliberalism as "market fundamentalism."[5] If this term is meant to imply a kind of fanaticism that worships the myth of the market, then the label is apt. But if it is meant to suggest that markets in a neoliberal age are in fact free — that government has simply vacated the field to let the market balance itself — then the label is demonstrably false. In our neoliberal age, all levels of government award massive subsidies to private, for-profit corporations, typically in the form of tax credits or grants. Government pays these corporations to influence what they do and where they do it. There is nothing free about this market. The amounts paid are enormous. Between 2000 and 2015, the federal government alone awarded $68 billion in grants and tax credits to influence corporate behavior in one way or another. Two-thirds of this total went to fewer than 600 companies. Six companies received $1 billion or more; 21 received $500 million or more; and 98 received $100 million or more. And this doesn't even count the value of subsidies awarded by state and local government. Neoliberalism is very big business.[6]

Neoliberalism does not work its mischief the same way everywhere. In policing and public schools, for instance, it encourages resources to shift to meet the needs of people with means at the expense of those without. In housing, it guarantees that people with the most capital will be given more, while those with the least are left to fend for themselves. In urban parks, it encourages the privatization of public space. But these differences are all symptoms of the same sickness. Everywhere that neoliberalism operates in American cities, it favors wealth over poverty. And where it cannot operate—where the private sector cannot find a way to make money from neighborhood institutions or the delivery of essential services—it demonizes government to ensure that these institutions and services are underfunded and overwhelmed. This creates a cruel gap: people in badly distressed neighborhoods like Olneyville desperately need investment and services, but the private sector ignores them unless the government guarantees a profit by providing massive subsidies.[7] Because the subsidies awarded by government are not remotely sufficient to make up for decades of disinvestment, distressed neighborhoods are chronically underserved.

The current approach to neighborhood well-being arose in response to this gap. All over the country, a small army of specialty nonprofit organizations, most of which operate on shoestring budgets, struggles to provide distressed neighborhoods with services that were once delivered by state, local, or federal government. Most of these nonprofits cannot survive without continual infusions of financial support from private foundations, philanthropies, and community institutions such as hospitals, universities, and local banks.[8] They operate a bewildering array of programs in the most distressed neighborhoods, from addiction counseling to Zumba classes. In several years of studying and writing about Olneyville, I have found these programs everywhere. The same partnership that established the Walking School Bus also created the computer education classes at the Olneyville branch of the public library and the financial literacy and after-school programs at D'Abate Elementary. A grant to Family Service of Rhode Island pays for social workers to accompany Providence police officers on calls that will likely prove especially traumatic, like the death of a child. A different grant allows a mobile food market to set up at D'Abate every Thursday evening to provide Olneyville residents with access to affordable, healthy, and culturally appropriate food. Other grants keep the doors open at a free clinic for residents who have no health insurance, relying on a rotating pool of volunteer physicians and medical students.

The nonprofit sector exploded as neoliberalism came to dominate American life. Per capita, the number of nonprofits in the country doubled between 1970 and 1990, and by 2000, the sector accounted for roughly 10 percent of U.S. employment and 8 percent of the domestic economy.[9] By 2015, more than 1.5 million nonprofit organizations were registered with the IRS. Of these, the largest number (more than a million) are public charities, a designation that includes most of the social service organizations operating in Olneyville. While the field is dominated by a small number of behemoths, the vast majority of public charities are tiny. Thirty percent report annual operating expenses of less than $100,000, and two-thirds report expenses less than $500,000.[10]

Almost no aspect of community life in Olneyville is not touched in some way by small, thinly funded nonprofits. One afternoon I was in the neighborhood, chatting with Meg Sullivan, the executive artistic director at the Manton Avenue Project. MAP is an after-school program for Olneyville schoolchildren who write their own plays under the mentorship of local artists. The plays are later professionally produced at a theater in Providence, with the volunteer help of adult directors, actors, and designers. Children can stay in the program for years and develop their creative talents and academic skills in a nurturing, collaborative environment. They also get the rare and memorable experience of seeing their work performed onstage. The program is free, as are the performances.

Like the Walking School Bus, the Manton Avenue Project receives funding from the Rhode Island Department of Health. We don't tend to think of an after-school writing and drama program as a public health initiative, but in a neighborhood like Olneyville, we should. In wealthier neighborhoods, parents spend considerable time and resources giving their children the chance to participate in after-school activities, and they view these programs as important to a child's emotional and thus physical health. Why shouldn't kids in Olneyville have the same opportunities?[11]

But children in Olneyville do not have the same worries as children in wealthier neighborhoods. Meg told me the kids in the program, like writers everywhere, tend to write about what matters most to them and their families. "Mostly they write about separation," about families ripped apart and parents stripped from their children. In 2018 separation was "a central part of their plays, with their characters facing mothers and fathers being taken away." The children in her program, Meg said, also write about being hungry. Every afternoon she asks her students to tell her one good thing and one "not so good thing" that happened that day. "Today was a good day," a young boy told her once. "Mommy went to the store."

On Friday afternoons, the school nurses at D'Abate hand out bags of food to many of the schoolchildren. For some families, the bags might provide nearly all the food they have for the week, especially at the end of the month. SNAP vouchers—what we used to call food stamps—are disbursed on the first day of the month, but they often do not last for thirty days. As one woman in Olneyville told me, "At the start of the month, the markets are crowded. At the end of the month, the food pantries are crowded." In any case, only citizens and a narrow category of foreign nationals are eligible to receive SNAP, meaning that many families in Olneyville could not get by without the food distributed by the school nurses. "Sometimes those bags are pretty heavy," Yuselly told me.

At its best, the current approach to neighborhood well-being helps ensure that local nonprofit organizations faithfully solicit the views of neighborhood residents. These organizations soften the effects of neoliberalism by bringing much-needed resources into a distressed neighborhood. If the nonprofits create enough initiatives like the Walking School Bus and the Manton Avenue Project, if those initiatives are adequately funded and intelligently administered for long enough, and if they are knit together into a comprehensive blanket of services, they can go a long way toward improving a place like Olneyville.

But however successful this approach might be, it will not protect Olneyville from gentrification, just as it has not protected hundreds of similar neighborhoods around the country. To begin with, and contrary to what is often imagined, neighborhoods like Olneyville are not funded. The money that comes into Olneyville does not go to the neighborhood itself. Instead, with only rare exceptions, the money goes to the State of Rhode Island, the City of Providence, or one of the many nonprofit organizations that work in the neighborhood. In 2017, for instance, the Rhode Island Foundation made $43 million in grants to 1,700 nonprofit organizations statewide that serve the neighborhoods where they are located. In late 2018, the foundation announced another $3.5 million in grants to six nonprofits "to reduce disease, promote health equity and address the medical and behavioral challenges of underserved populations."[12] The Rhode Island chapter of the Local Initiatives Support Corporation (LISC) contributed another $455,000 in 2017 to seven community development corporations.[13]

Community banks, universities, hospitals, and private businesses also make grants to benefit Rhode Island neighborhoods, but once again the recipients are typically nonprofits. Bank Newport, for instance, reports that its grants "are generally limited to nonprofit organizations that have a bank and/or insurance relationship with us." Then there are the block grants made by the federal government, including the Department of Housing and Urban Development's Community Development Block Grant Program, which awarded $3 billion in grants in 2019, all of which went to eligible cities and counties. (Though this sounds like a lot of money, after adjusting for inflation, CDBG funding has decreased more than 80 percent from its peak in 1979, and between 1980 and 2015, the average grant for each recipient fell 85 percent.)[14] In short, neighborhood residents depend on what outsiders give them. They do not own or control any of the money earmarked and spent for their benefit and cannot invest it for the long term.

Because residents do not control the assets that flow into Olneyville, agencies and actors who are distant from the neighborhood control whether programs like the Walking School Bus will continue. If any of the links in a long chain were cut—if the CDC cut its support, if RIDOH shifted its funding priorities, if ONE Neighborhood decided to take its work in another direction—the bus would stop running. This lack of control puts the community perennially at the mercy of outside funders. And that funding is rarely sufficient and always at risk. The authors of a report by the National Research Council and Institute of Medicine found that funding for youth programs was "modest compared with the number of children who need assistance.... If there is one barrier above all others to an ample supply of high-quality community programs for children whose parents cannot afford to pay for them, it is the lack of reliable, stable funding streams to support them."[15]

What is true for youth programs is true for nearly every aspect of neighborhood life. It is almost impossible to overstate how precarious life is for many of Olneyville's residents, who are chronically close to homelessness and ruin. In the entire neighborhood of about seven thousand people, where one in three lives below the poverty line (compared to the national rate of one in seven) and 20 percent of households survive on $10,000 per year, there is precisely one food pantry.[16] The staff there told me it serves about seventy clients per day. Clients can come in twice a month and are given a carefully controlled allotment of food based on their family circumstances. The pantry is open four hours a day, four days a week. For want of funding, it closes on Friday at 1:00 p.m. and does not reopen until Tuesday at 9:00 a.m.

I spoke with Jackie Reyes, who works at the pantry. She came to this country as a child from the Dominican Republic and lived for many years in New York on Eighty-Fifth and Broadway before moving to Providence. Trained as a chef, Reyes says her personal goal is "to feed people." Yet she also understands that no one who comes to the pantry is just hungry. They have other, overlapping needs. For that reason, she wants the pantry to be a place where people can learn about other resources available to them. "I want it to be a hub," she said, "like Grand Central Station, where people can go to different places. It's all connected." She nodded to a corner of the pantry just inside the door, where brochures in English and Spanish advertise childcare, wellness workshops, and assistance with SNAP benefits and utilities. She wants to help people "get up from the bottom," but she knows how hard this is. I asked her what happens if a family runs out of food on a Friday afternoon, knowing the pantry will not reopen until Tuesday. She looked at me and shrugged, as if to say, "Welcome to Olneyville."

Just in the time I was researching and writing this book, an important nonprofit organization that had been a fixture in the neighborhood for years lost its funding. English for Action taught language classes to Olneyville residents, at one point teaching more than a hundred students per semester. But the organization did much more than teach English. It linked the development of language skills to community building and tried to give Olneyville residents the tools they needed to become leaders in the neighborhood and advocates for their neighbors. It was one of the only organizations in Olneyville to adopt a consciously radical perspective, working to "challenge existing power dynamics and [the] distribution of resources in our society" and conceiving itself as "comrades in movements that advance immigrant rights."[17] And unlike most nonprofits in the neighborhood, English for Action made a deliberate effort to build its staff by hiring, training, and promoting neighborhood residents. Yet despite its importance and success, English for Action depended wholly on outside support. When the Rhode Island Department of Education cut the funding, the organization quickly died. It closed its doors on December 31, 2018.

But even if funding streams in Olneyville were adequate and

secure, there is—and ought to be—an inherent preference for controlling one's own fate rather than having it controlled by others. This is not just a political necessity but a matter of historical experience. In a way that I did not appreciate when I started my research, change in an urban neighborhood is inevitable. A city is not so much a fixed point on a map as a performance, an endless work in progress. A city is the immensely complex product of countless relationships among a constantly changing cast of residents, businesses, and visitors, who continually reshape the place to fit their needs.

This change has never been an agentless process. A neighborhood can either act or be acted on, and if low-income residents do not take charge of their fate, others will. Outsiders will decide how the space should be used and what is best for the people who live and work there. Outsiders will decide whether to build more housing, and whether that housing will be affordable. They will decide whether to invest in education and public health, and what those investments might look like. They will decide how the neighborhood is policed. More often than not, when others decide what is best for a low-income community—especially a community of color—they do not make it better. When outsiders set the priorities, low-income Black and Brown neighborhoods have too much of what they don't need and too little of what they do. They do not have safe, healthy, affordable neighborhoods. Instead they have neighborhoods that are underserved and overpoliced. An outside organization may be an excellent and benevolent steward, but the low-income residents of a distressed neighborhood do not need a steward to run their lives for them. They need to run their own lives.[18]

Finally, because outside funders and agencies are often limited in the approaches they can take, either by law, charter, or tradition, many of the initiatives established under the current model do little to protect low-income neighborhoods from the economic and political forces that threaten to destroy it by marginalizing and eventually displacing people of limited means. The very initiatives that make the neighborhood a better place to live can make it irresistible to the back-to-the-city movement and its attendant influx of unregulated capital. Unless guardrails are put in place, this capital will ultimately change the character of the neighborhood and raise the cost of essential goods like rent, property taxes, health care, insurance, and food, making the neighborhood more difficult and less desirable for the poor to live in. Perversely, transforming the neighborhood can lead to its destruction as a viable home for low-income residents. And in the final indignity, the people displaced are frequently Black and Brown, while those who displace them are typically white and comparatively wealthy, intensifying both the racial and class stratification of American life.[19]

Therein lies the moral obscenity that motivates this book: On one hand, the current approach is neither adequately funded nor locally controlled and therefore cannot achieve its full potential to transform a distressed neighborhood. On the other hand, when against all odds it makes a neighborhood better—as it has in Olneyville—its success accelerates the gentrification that eventually makes the neighborhood unaffordable. This paradox has too often gone unexamined. Much of the writing about urban well-being amounts to instruction booklets for tweaking the current approach. But the effort may ultimately be self-defeating, since the success of the model can lead to the failure of the neighborhood. To protect Olneyville, we need more than well-oiled nonprofits. We need something that is funded, controlled, and organized from within the neighborhood.

That something is a neighborhood trust, which I describe later in the book. For now, it is enough to note that the trust is a legally protected way for residents to pool the money that comes into their neighborhood, and use it to place land and nonprofit organizations under communal ownership and control. But all in due course.

Countless neighborhoods have traveled the path from working-class vibrancy to postindustrial blight, but the passage always leaves a distinct mark, and no two neighborhoods emerge with identical scars. At the beginning of the twenty-first century, the most pressing problems in Olneyville revolved around the nearly complete collapse of the physical infrastructure; centuries of environmental pollution; an absence of quality housing; high rates of crime, disorder, and police brutality; and a lack of anchor institutions to unite the neighborhood. Olneyville faced other problems as well, but those were the most critical, and as I describe the neighborhood's transformation, they are the topics I cover.

Other neighborhoods might face different problems. Elsewhere, for instance, the most pressing challenges could be a lack of public transportation and a legacy of racial violence and discrimination. In still others, the population might be primarily native-born Black rather than immigrant Latino, which could produce a different relationship with authority. Describing the transformation of those neighborhoods would be a different book, but not radically different. The problems confronting distressed urban neighborhoods in this country are merely different symptoms of the same underlying sickness. They are the product of the historic, economic, and political processes that swept through urban America in the second half of the twentieth century.[20]

All these neighborhoods now operate in the same neoliberal environment, where the private sector is imagined as the magical solution to any problem worth solving. In some places, neoliberal thinking maintains a tighter grip than in others, but it is hegemonic almost everywhere in the United States and has broad bipartisan support. To meet the challenges before them, distressed neighborhoods all over the country field the same weary foot soldiers: a small army of chronically underfunded nonprofit organizations, never more than a grant cycle away from closing their doors, that must scramble year in and year out for enough cash to operate tiny, hyperlocal programs like the Walking School Bus, the Manton Avenue Project, and English for Action. And all these places face the same existential question: Can the neighborhood be transformed but not destroyed?

In answering this question, we find two conflicts simmering just beneath the surface. The first is the long-running tension between neighborhood and capital. The nonprofits that work in Olneyville have had great success in transforming the neighborhood. Unintentionally, they have made it attractive to unrestrained capital, which increasingly turns its attention to the small west-side neighborhood it long ignored. After decades of neglect, rents and housing prices are accelerating rapidly, and new businesses are opening.[21] The city has encouraged this trend, using the neoliberal tools at hand to lure investors. In 2018, for instance, the entire neighborhood was designated an Opportunity Zone—a creature of the 2017 Trump tax cut—creating powerful financial incentives for developers to invest in the neighborhood, but extracting no guarantee that those developments will produce more than a tax break for wealthy investors, rather than a benefit for low-income residents.

Can Olneyville benefit from increasing investment without embracing gentrification? We cannot afford a negative answer to this question. Cities have historically been the place where the deck of American life is most reliably cut and shuffled. They are the site of expanded horizons, enlarged opportunities, and enriched lives. Yet that is rapidly changing. Cities across the country are becoming increasingly unaffordable and driving out the working poor. Between 2000 and 2015, almost every major metropolitan area in the country saw poverty rates soar in the suburbs. In Sunbelt cities like Austin and Las Vegas, the number of people living below the poverty line in the suburbs more than doubled; in the suburbs surrounding midwestern industrial cities such as Chicago and Detroit, the number climbed by more than 80 percent. Even in thriving regional centers such as Washington, D.C., and Seattle, the number of poor people in the suburbs increased by more than 60 percent.[22]

As the working poor move to the suburbs, the cities they leave behind cleave into unaffordable enclaves set alongside pockets of concentrated distress. Since 2000, most of the large metropolitan areas in the country have seen a steep increase in the number of census tracts with a poverty rate above 40 percent. By 2014, despite the economic rebound, 14 million people lived in these extremely poor neighborhoods, twice as many as in 2000. In such neighborhoods, almost everything that matters to a person's future—from levels of crime and disorder to the performance of schools; from the quality of the housing to the opportunities for social mobility; from physical and mental health outcomes to the prevalence of severe financial insecurity—tends to be worse. The longer a person lives there, the lower the chance of ever escaping poverty.[23]

When cities become isolated islands of rich and poor, residents lose the countless formal and informal encounters that promote tolerance in a healthy, safe, and well-integrated city. At least since the early 1960s, when the sociologist Jane Jacobs wrote *The Death and Life of Great American Cities,* we have recognized this neighborhood effect as crucial for producing the economic growth and innovation for which successful cities are renowned, as well as a society that is tolerant of differences.[24] Yet we are steadily losing the social and professional interactions that occur when a diverse mix of people live, work, worship, and play in close proximity to one another.

The second conflict is less obvious but equally important. It is the tension between progressive elites and local self-control. By and large, the organizations I studied in Olneyville are staffed by dedicated, well-educated whites with an abiding commitment to the neighborhood and its seven thousand residents. They share a progressive vision for Olneyville and want nothing more than to help neighborhood residents overcome the austerity imposed by a neoliberal age. But ultimately their voice is not the voice of the low-income Black and Brown residents.

The residents of Olneyville are perfectly able to speak for themselves. They can decide what is best for them and their neighborhood. Implicit in this is the recognition that, if allowed to choose, the residents may not follow a progressive path. While they might follow the course laid out for them by well-intended nonprofits, or by me in this book, they might venture in an entirely different direction. What matters is not the path they follow but that they have charted it themselves; such is the essence of democracy. That is why I do not purport to speak for the residents of Olneyville in this book. I interviewed many current and former residents during my research, and their voices appear throughout these pages, but I do *not* claim to speak for them or to tell them what to do. My goal is to empower them so that they might chart their own future.

To be poor in this country has never been easy, and it is getting harder every day. As we open the third decade of the twenty-first century, the attacks on the poor, especially poor people of color, are as vicious as they have been in many years. Three weeks before Christmas 2019, the Trump administration approved changes to the food stamp program that are expected to remove 700,000 people from the rolls.[25] Neoliberalism is ascendant, social welfare is in retreat, racism is emboldened, and inequality is at record levels. Yet this book offers hope. Rather than lament what is, it imagines what can be. This hope is guided by the radical idea that *everyone* can live in a safe, healthy, vibrant, and affordable neighborhood. The rich can buy it. The poor must build it. Olneyville can show us how.

2

"Not a Particle of the Romantic"

For most of its long history, Olneyville has been what it is today: a neighborhood of low-income hardworking immigrants. For generations, the people who have called Olneyville home have chiseled out a precarious toehold to survive in chronically low-wage employment. In good times, factory owners provided jobs, but the pay was bad and the work was dull. In bad times, the owners simply sent workers home without pay until demand picked up again. When other regions beckoned with a promise that workers could live on less, owners packed their bags and moved away, desolating the neighborhood. Throughout this history, the political class in Providence has been an unreliable partner, sometimes protecting Olneyville, other times doing it harm, but mostly leaving it to fend for itself. Today, when we ask whether the neighborhood can be transformed but not destroyed, we must bear this history in mind. Nothing in Olneyville's past justifies the view that unrestricted capital or the political elite will save the day.

The magnitude of Providence's fall is almost impossible to conceive. In 1900, the city ranked among the ten largest industrial centers in the country. It was home to the world's largest manufactories of tools (Brown & Sharpe), files (Nicholson File), engines (Corliss Steam Engine Company), screws (American Screw), and silverware (Gorham Manufacturing Company). Providence also led the United States in jewelry manufacture: nearly seven thousand artisans made jewelry at more than two hundred firms citywide.[1] Yet within a single lifetime, the city would become a half-empty hulk, mottled with abandoned plants, shuttered storefronts, vacant properties, and collapsing houses.

Textiles were the early king of the Providence economy. Unlike some other parts of New England, Providence made most of its fortune on wool rather than cotton. In 1890 it produced more woolen and worsted goods than any other American city.[2] The center of the Providence textile industry was Olneyville. Like much of Rhode Island, Olneyville throws its roots deep into precolonial history. Native Americans settled the area long before Roger Williams purchased the land from the Narragansett Indians in 1636. White settlement in the area began early in the eighteenth century along the banks of the Woonasquatucket River. By the mid-1700s, the area had a paper mill and distillery, as well as a road built in 1745 that survives today as Valley Street. Additional light industry followed, including a grist and paper mill operated by the Revolutionary War veteran Christopher Olney, for whom the neighborhood is named.[3]

Textiles arrived in the early 1800s, and for much of the century, Olneyville was an independent mill village. The oldest surviving industrial buildings in the neighborhood date from the 1830s, but the true giants of the industry arrived in the half century afterward. The Providence Dyeing, Bleaching and Calendering Company, which finished and prepared textiles for the retail market, opened its plant on Valley Street in 1846.[4] Most of the complex remains standing; it now holds, among other businesses, the coffee shop where I wrote much of this book. Atlantic Mills, which would become the neighborhood's largest employer, began operations in 1851. It too remains, nestled between the Woonasquatucket River and Manton Avenue, one of the three major streets in Olneyville.[5]

The Riverside Worsted Mills, a sprawling complex between the

Woonasquatucket and Aleppo Street, opened not long after the first shots of the Civil War and quickly saw heavy demand for coffin covers, which "cost very little" but "sold for good money."[6] In the 1880s, the Providence and National Worsted Mills, the "giant of the Providence woolen industry," opened on Valley Street. In an 1890 business survey, city fathers described Providence and National, in the extravagant boosterism of the day, as "the most extensive single plant in the world devoted to the production of worsted yarns and worsted goods."[7] On the site today stands Rising Sun Mills, a campus of residential apartments alongside corporate and professional offices. The apartments are named for the paper mill built by Christopher Olney before the Revolutionary War.

Mill owners built their plants close to the major thoroughfares that encircled the neighborhood—Manton Avenue, Atwells Avenue, and Valley Street—so that they would have ready access to transportation in and out of the city. The mills surrounded Olneyville's residential core, a patchwork of modest, wood-sided single- and two-family houses along with a few of the triple-deckers common in New England. This arrangement allowed most residents to walk to work in one of the nearby factories.

A streetcar started operating in Providence at the end of the nineteenth century, with the line from downtown stopping at Olneyville Square, a boisterous jumble of a half-dozen streets at the eastern edge of the neighborhood. The advertisements in the weekly *Olneyville Times* tell us something of the area's vitality. In and around the square, shoppers could get a new horse carriage; stock up on coal, hay, straw, and grain; buy $10 suits and overcoats; pick up a new hat or a pair of handmade boots or shoes; and get the latest cure for malaria. In 1897, W. D. Harris, a cash grocer on Plainfield Street, proclaimed, "All customers treated alike; one man's money is as good as another."[8] In the next year, Providence annexed the bustling mill town, and Olneyville became part of the city. Olneyville Square became its second-largest commercial district and was known as "the second downtown."[9]

Bustle should not be confused with prosperity. The economics of the textile industry have never been kind to workers. Most work in the mills was labor-intensive, low skilled, and repetitive. Textile workers were, on average, among the lowest paid in the entire manufacturing sector throughout the nineteenth and twentieth centuries. One study, from 1955, found that average wages in the textile industry ranked thirty-second of thirty-three industries, with hourly wages less than half those of the construction industry, the top-ranking sector.[10] At the same time, the biggest expenses in the textile industry were raw materials and labor. The mill owner could not control the cost of cotton or wool, so his profit depended largely on his ability to suppress labor costs, inevitably pitting workers against owners.[11]

Because a small textile mill could be just as efficient as a large one, the textile industry did not experience the widespread consolidation seen in other industries. This led to a productive capacity that far exceeded demand, depressing prices and putting additional pressure on wages. Production became exquisitely sensitive to fluctuations in demand, which made work extremely precarious. Workers grew accustomed to the fits and starts of the demand cycle. One week they might have full-time work, the next a fraction of that, the week after none at all.[12] This was before social security, unemployment compensation, workers' compensation, and minimum-wage laws.

Then there was the legendary tedium of work in the mills. As a writer described it in 1886,

Life in a factory where any textile industry is carried on is perhaps, with the exception of prison life, the most monotonous life a human being can live.... [A weaver] has got at least six looms to tend. They are arranged in a double row and his position is between them. He passes from one to the other. He must keep his eyes on them all and be ready to "change the shuttle" when the "filling runs out." He tramps thus back and forth up and down his "alley" for five hours, with no time to sit down and rest of a moment. After dinner he resumes his position at the looms and repeats the story and this goes on day after day, week after week, for months and years, the same round of toil, with little or no change from year end to year end, realizing even by this unremitting toil only just sufficient to feed and clothe himself and his family, and however desirous he may be to save a little sum for a rainy day he finds himself unable to do so. This is the lot of the weaver. There is not a particle of the romantic in his life.[13]

Because most mill work was poorly paid, dull, and irregular, it went to those who had the fewest other options, which in the United States has always meant the poor and nearly always meant immigrants. In the decade after 1845, when the Great Famine struck Ireland, more than a million Irish streamed into the United States. They quickly replaced native-born workers in the mills in Olneyville and elsewhere. In the 1860s, an agricultural crisis in Quebec unleashed a flood of Quebecois into New England—200,000 by 1870 and half a million by the end of the century—many of whom likewise found work in the Olneyville mills. Between 1880 and 1924, four million Italians fled to the United States from the grinding poverty in southern Italy and Sicily, making Italians and Irish the largest immigrant communities in Providence. Then came Portuguese, Poles, and Russian and Slavic Jews. Wave after wave of newcomers settled in Olneyville and neighborhoods like it, hoping—like immigrants since time immemorial—that new would also be better. By 1915, two-thirds of the city's residents were either immigrants or their children. For decades, the streets and shops in Olneyville were filled with the sounds of Gaelic, French, Italian, Portuguese, Polish, and Yiddish.[14]

Even in the best of times, life in a northern textile town was hard. In the twentieth century, it got harder. After the Civil War, southern politicians and business elites joined to woo the cotton textile industry from its New England home. "We felt no danger from the South until 1880," one textile executive said. "Then the cloud was no bigger than a man's hand, but it was there and was threatening us."[15] The champions of the New South promised New England industrialists cheap land and a ready supply of low-paid workers. In the late nineteenth century, the prevailing wage for textile workers in the South was about half what it was in the North. In addition, southern leaders assured northern mill owners they would not be bothered by the annoying labor movement that cut into profits in New England by restricting hours, closing the factory doors to children, and regulating working conditions. The native-born workers in the South, they promised, were more pliable and obedient than the fractious immigrants in New England. Moving south would end labor strife.[16]

Decline came slowly, and some parts of New England maintained their dominance longer than others. Because Providence mills mostly spun wool rather than cotton, they were less directly threatened.[17] Yet Providence exists within a regional economy and could not isolate itself from the wider decline. At first, New England mill owners tested the southern waters by opening single, smaller subsidiaries in the South. But as the competitive advantages became clear, especially in wages, and as southern mill hands became more skilled, New England owners transferred more and more of their operations to their southern plants, which they equipped with the most efficient, labor-saving machinery.[18] By 1915, textile production in the South had nearly caught up with that in the North. Both the North and

South expanded production dramatically during and immediately after World War I, but by 1925, capacity in the North was plummeting. In 1930, the total value of cotton textiles produced in the South was more than twice that produced in the North.[19]

The Great Depression and the commercial development of synthetic fibers hastened the industry's decline. In 1932, the GDP fell by 12.9 percent, the steepest single-year decline of the twentieth century. In the next year, the national unemployment rate hit nearly 25 percent, the highest level ever recorded.[20] A year after that, average earnings in the textile industry were roughly 40 percent lower than the average for all other manufacturing workers, and Olneyville was suffering badly.[21] The Federal Home Loan Bank Board described the housing stock in the neighborhood as "characterized by age, obsolescence, and ... infiltration of a lower-grade population," and therefore a bad risk for mortgage lenders.[22] In 1936, Riverside Mills, which had employed more than two thousand workers in the 1880s, closed its doors.[23] In 1941, the Work Projects Administration released an assessment of housing in the city and found that nearly every block in Olneyville was either "blighted" (30 to 50 percent of the housing substandard) or "slum" (over 50 percent substandard).[24]

World War II provided a temporary shot in the arm. But the defense industry that took shape during the war was the first casualty of peace. The massive Walsh-Kaiser shipyard at Fields Point in Providence, which employed more than twenty thousand people at its peak in 1944, shuttered almost as soon as the guns fell silent, and by mid-1945 it employed fewer than four hundred workers.[25] By the early 1950s, the cotton textile industry of New England was a lost cause; southern mills employed eleven out of every twelve cotton mill workers. The challenge now was to preserve the woolen and worsted trade, where New England still controlled two-thirds of the market.[26] But in 1952, the woolen and worsted industry had its worst year since the recession of 1938. An emerging consumer preference for lightweight, casual clothing made from synthetics depressed the demand for wool, and excess postwar capacity kept prices low. Mill owners responded by cutting wages, reducing hours, and trimming the workforce, and Olneyville, long teetering on the edge, finally fell over the precipice.

In 1952, an article in the inaugural issue of *Challenge* magazine decried the woeful state of New England's textile industry. The author pointed to a few New England plants, including the massive Atlantic Mills plant in Olneyville, where owners had modernized their operations, and mused hopefully that this greater efficiency might offset the South's considerable advantages.[27] But it was not to be. In the same year the article appeared, the Providence Dyeing, Bleaching and Calendering Company shuttered after more than a century in operation. In 1953, Atlantic Mills—at one time the largest employer in the neighborhood—closed its doors. The American Woolen Company, a conglomerate that owned dozens of mills throughout New England, including several in Olneyville, shut down all its remaining plants during the 1950s. Between 1953 and 1957 alone, jobs in the already depressed textile industry across Rhode Island fell by a third.[28]

Textiles left Providence first, but the other industrial giants were not long behind. American Screw left for Connecticut in 1949. A decade later, Nicholson File transferred its manufacturing operations to Indiana and its administrative and sales department to East Providence.[29] Brown & Sharpe decamped for the Providence suburbs in 1964. In 1975, U.S. Rubber closed its plant on Valley Street, just north of the Olneyville border.[30] Sometimes a company sold itself to a foreign buyer, which instituted another round of cost cutting as demand continued to fall. But most of these efforts merely delayed the inevitable. One after another, nearly all the giants of Providence's manufacturing age closed their doors.

The city fell into a postwar abyss. Between 1940 and 1950, the population fell by only 2 percent. But over the next decade, as the suburbs grew, Providence lost 16 percent of its population. By 1970, the city had shrunk to fewer than 180,000 residents, a two-decade decline of nearly 30 percent—the largest proportionate population loss of any major city in the country.[31] A decade later, the city's population was just over 156,000, a decline of over 40 percent from its 1925 peak.[32] Olneyville's decline was even more precipitous. Between 1950 and 1965, in a neighborhood that covered just half a square mile, at least six major plants closed their doors, and between 1950 and 1980 the population fell by half.[33] In 1983, the Providence Redevelopment Agency reported that the "exodus" from the neighborhood "has left a population that is, while still blue-collar, of a much lower income level than in the past."[34]

As the city emptied, the suburbs filled. As elsewhere, the lure of suburbia in Providence was almost mystical. The suburbs represented a clean, safe, sky-blue nirvana, untroubled by the blight, ugliness, and despair that had engulfed so many U.S. cities. The Pratt Attitude Survey of Providence residents in 1956 revealed "a compulsive urge among residents of the city to move to the suburbs or the country" in search of "a better life." More than half of those surveyed had considered moving out of the city, and of those, three in four thought they actually would. "The automobile and the desire for light, space, and accessibility are moving people outward—stores and industries are following."[35]

But this vision of the good life had a strong push from government. Low-cost mortgages from the Federal Housing Administration meant that for people who could get a mortgage, buying a new house in the suburbs was cheaper than renting a comparable apartment in the city. In Providence and elsewhere, those who could afford to move fled inner-city neighborhoods like Olneyville for new single-family homes in the suburbs.[36] As Providence collapsed, the nearby suburbs boomed. Between 1940 and 1980, the combined population in Cranston and Warwick more than doubled, from 76,000 to 160,000.[37]

But what good are a brand-new home and a shiny new car without a highway to get you where you need to be, quickly and painlessly? People who moved to the rapidly expanding suburbs demanded smooth and fast-flowing arteries that would allow them to travel quickly and easily back into the city and, in time, from suburb to suburb. The Truman administration resisted a national highway program, but the Eisenhower administration embraced it enthusiastically. In 1956, Congress passed the Highway Act, which provided for the creation of a national highway system of forty-one thousand miles. The federal government promised to pick up 90 percent of the tab.[38]

Providence's first expressway was deliberately designed to circumvent Olneyville so that drivers could avoid the rapidly declining Olneyville Square. Construction started in 1949, and the Olneyville Bypass opened in 1952, just as the woolen and worsted industry was set to fall over the edge.[39] You can't build a highway through a major city without buying up the land, kicking out the current owners, and bulldozing their businesses and homes. All told, the city seized more than three hundred parcels of land in Olneyville, and the list of the displaced reads like a roster of urban ethnic America.[40] Antonio and Joseph De Vincenzo lost the home they owned on Broadway, as did Nicola Bergantini and his wife Angelina. Eden Street was completely destroyed, taking with it the homes of Manzo Costanzo, Vito Benedetto and his wife Rafaella, and Alexander Chmielewski. The German Dramatic Society of Olneyville lost the hall it owned on Bell Street, and Henry and Celia Englund were kicked out of their house on Harris Avenue.[41] The sunken highway encircled half the neighborhood like a moat, becoming Olneyville's southern and eastern border and cutting it off, physically and psychologically, from the rest of the city. The streetcar from downtown had stopped running in 1948, and Olneyville became increasingly isolated.

Like any major redevelopment project, the highway plan was sold with the cheerful promise that the temporary disruption caused by the construction would make everyone's life better in the long run.[42] It never did. As jobs and people left, the neighborhood started to show the unmistakable signs of abandonment. In 1975, the Providence Planning Department conducted a visual inspection of the houses in Olneyville and found that fewer than one in four remained in excellent condition. Just three years later, when they repeated the inspection, the number had fallen to fewer than one in twenty. Over half the houses in Olneyville needed "immediate attention," were in a state of "advanced deterioration," or were "heavily deteriorated and dilapidated." The department described Olneyville as a "run-down," "low-income" neighborhood that faced "an uphill struggle" to reclaim "its former prosperity and liveliness."[43]

The department resorted to the comically euphemistic language of bureaucrats everywhere to describe the neighborhood's virtually nonexistent economy. Olneyville Square, once the commercial center of the neighborhood, was "waiting for revitalization," which is a kind way to describe boarded-up storefronts and abandoned shops. In the same vein, the department took note of the "large number of vacant factories . . . waiting to be renovated."[44] But any implication that a queue had formed, and Olneyville was patiently waiting for its turn in the revitalizing sun, would have drawn a quizzical look from residents who daily passed by the idle plants and empty stores. Olneyville had already been waiting a long time.

Vacant lots and crumbling homes are just two of the many indicators of a city on the ropes. A city lives on its tax base, and a declining city has a shrinking pie, with smaller slices for everything the city needs to buy and everyone it needs to pay. Roads go unpaved, trash uncollected, streetlamps unlit. Teachers and firefighters are laid off. Public hospitals can't buy equipment, police departments can't respond to calls, and the city ceases to function. The impact is even more dramatic at the neighborhood level. When a neighborhood grows thick with boarded-up properties and overgrown space—when half the homes are empty and the other half are falling down; when the sidewalks buckle and the streets aren't safe; when shops close and jobs disappear—the few people who remain tend to withdraw. The dense web of social networks and personal relationships that once tied a healthy neighborhood together begins to fray and ultimately rip apart.[45]

Being neighborly takes work. It takes time and energy to gather with neighbors, join clubs and committees, attend meetings, and press to make the neighborhood better. It takes effort and no small amount of courage to admonish someone's children not to spray graffiti on the wall, smash bottles in the street, or sell drugs on the corner. And people are least apt to invest that time if it seems pointless, as it would in a neighborhood that has been neglected for decades. When a city is collapsing, nothing seems more obvious than the futility of collective action. In 1975, Providence mayor Buddy Cianci came to Olneyville and implored attendees at a neighborhood meeting to "look at the facts, one of which is that Providence is a tired old city."[46] His declaration could not have been news to anyone in Olneyville, where municipal decrepitude had been obvious for years.

Neighborhoods that reach this condition have lost what academics call social capital, meaning the collective value of the social networks within a community. Markers of social capital include things like trust in one's neighbors, participation in community organizations, informal mutual assistance, and effective collective action.

When these markers decline, so does the quality of life. When residents do not trust their neighbors, when they withdraw from community groups, and when they decide that collective effort is not worth the trouble, the neighborhood has lost more than jobs and municipal services. It has lost the sense of community and mutual obligation that distinguishes a healthy, vibrant, safe neighborhood from a sea of anomie and despair.[47] As jobs disappear and neighbors leave and are not replaced, more properties are abandoned, creating havens for illicit activity and destroying the sense of safety that comes from having benevolent eyes on the street. Just as no one could fault residents who have options elsewhere for pulling up stakes and moving away, no one can blame those who are trapped in a disintegrating neighborhood for collapsing into hopelessness.[48]

Yet by the twenty-first century, this "tired old city" was known not for its long decline but for its hip downtown. One journalist called Providence "an economic and cultural force." Another said it was "a national model for how to make a run-down old city hot again." *Money* magazine, touting the city's "remarkable rebirth," dubbed it the best place to live in the Northeast, and *Travel and Leisure*, celebrating Providence's "edgy galleries" and "sophisticated dining," called it "New England's most exciting city" and the third most charming city in the country after New Orleans and Charleston.[49]

Providence came late to downtown revitalization. For the first three decades after World War II, the governing coalition in the city remained passive as one business after another closed or moved to the suburbs and tens of thousands of residents left the city. It was not until the mid-1970s that Providence, far behind other cities, finally intervened. But once it did so, it followed a path that by then had become familiar. A coalition of local business leaders provided the vision and elite support, state and local politicians acquired the land and swept aside regulatory hurdles, and the federal government covered most of the cost. Providence refashioned its downtown around the familiar leaders of the new economy: financial services, retail, and tourism. The city leveraged federal money to relocate the train tracks and redirect the course of the Moshassuck and Woonasquatucket Rivers, freeing up virgin land that was subsequently given over to the tourist and retail projects that included a conference center, a shopping mall, high-end hotels, luxury apartments, and headquarters for a bank.[50]

The revitalization of downtown Providence thus looked much like the revitalization of many urban centers across the country: the federal government paid billions of dollars to facilitate a private development plan conceived by local elites and facilitated by state and local government. But as elsewhere, this churning had uneven results. As careful students of the Providence transformation have observed, "the kinds of jobs created by all this activity are of two main types: information and technology-intensive jobs that reward a well-educated population, and low-order retail and service positions that provide a weak foundation for widespread prosperity."[51] The service jobs created by the city's downtown "renaissance" had neither the economic clout nor the security of the manufacturing jobs they replaced. "New jobs have been created," an official with the Rhode Island AFL-CIO acknowledged in 1985, "but we're finding that the people who have been laid off or lost jobs because of a plant closing, when they find a job it's often paying $2, or $3, or $4 an hour less than they had been earning. And then there's benefits, which are often overlooked."[52] As a result, city residents "thus far have little to show for all the activity beyond municipal property tax rates and a per capita public debt that rank among the nation's highest." For the residents of neighborhoods like Olneyville, an "extraordinary redevelopment project" produced "what are likely to be rather ordinary outcomes."[53]

The oligarchy that drove this transformation forward took little heed of the struggling neighborhoods on the city's perimeter, which continued to decline as the center revived. Crack cocaine swept through Providence beginning in the mid-1980s. In 1987, the mayor complained that despite arresting "pushers" at "record" rates, the Providence police were losing the war on drugs. In the next year, a police lieutenant lamented that the state's drug problem was "almost reaching epidemic proportions."[54] Olneyville, already one of the poorest and most depressed areas in the city, was hit hard. In February 1989, Mayor Joseph Paolino promoted the *Newsweek* cover story that named Providence one of the country's "hot cities," the only city in New England to make the list. But while the mayor gushed, *Ocean State Business* was reporting a "crescent of instability, dominated by low-income residents, worn-out housing and multiple social problems extending from Olneyville to South Providence."[55]

In June 1989, the chief of police led more than three dozen uniformed officers in a "spring offensive on drugs" in Olneyville and another neighborhood, arresting more than 125 people in a single night. But the next month, residents met with state and local officials at Saint Teresa's Church in Olneyville to complain that drug dealing was rampant in the neighborhood. As the meeting took place, an undercover officer was buying cocaine in an Olneyville bar five blocks away. And the next year, an Olneyville resident complained to the police that drug dealers were "terrorizing" the neighborhood.[56] Between 1974 and 2000, while Providence was becoming what *GQ* magazine called "the coolest city you haven't been to yet," the murder rate in the city more than doubled, rapes increased fivefold, and robberies climbed by 20 percent.[57] In 2002, Olneyville had the third-highest rate of violent crime—meaning murder, rape, robbery, and aggravated assault—of the city's twenty-five neighborhoods.[58] Olneyville acquired a reputation around town as the place that never caught a break. "It was like there was always a cloud hanging over it," a life-long Providence resident told me one evening. "We used to call it Lonelyville."

The 1980s brought two changes that continue to shape the neighborhood today. First, as the ideology of neoliberalism ascended, the federal government dramatically reduced its support for cities. During Ronald Reagan's presidency, direct and indirect federal aid to cities declined by 60 percent.[59] Since the 1980s, the federal government has continued to reduce support for social welfare. Between 1978 and 2016, for instance, aid awarded by the Community Development Block Grant Program, the largest city-aid program, fell 80 percent, from $12.7 billion to $3 billion in 2016 dollars, and the Trump administration threatened to stop these grants altogether. Eliminating the program would strip funding from nearly 1,300 cities and towns that rely on the money to develop housing at below-market rates, repair basic infrastructure, provide job training and youth programming, and address countless other local needs.[60]

Nor could cities count on state legislatures to make up the deficit. During the Great Recession, most states, including Rhode Island, cut aid to cities. A decade after the financial crisis, when many state governments had not yet fully recovered, Rhode Island was still spending fewer dollars on aid than it had in 2008.[61] State aid to local governments nationwide hit a postrecession low in 2013, when it was down by 5.3 percent. Rhode Island was worse; its state aid to cities decreased by 11.4 percent after the recession. For local governments, state aid is the second-largest source of revenue after tax dollars, so these cuts drastically pare city budgets and impose real harm on the lives of residents.[62] In Central Falls, a few miles from Providence, state-imposed austerity drove the city into receivership, after which the receiver appointed by the state cut retiree benefits by 55 percent.

In Providence, the budget shortfall forced the city to reduce benefits for municipal workers.[63]

To replace this lost revenue, cities now try to lure the high-wage firms of the knowledge economy. They do this in a number of ways, but the most important is to promise generous tax breaks. From 2017 to 2019 alone, state and local governments lost nearly $45 billion in revenue as a result of tax abatements given to private companies.[64] Yet these businesses are precisely the sort that reinforce the structural inequity of the hourglass economy. Cities are thus caught in a trap: they need the knowledge economy to fund their operations, even though the knowledge economy exacerbates the inequality they hope to reduce.

The second major change that began in the 1980s was demographic; Olneyville gradually turned from white to Brown. For most of its long history, Providence was overwhelmingly white. In 1950, Blacks represented only 3 percent of the city population, and Hispanics (as the Census Bureau then described them) did not even register in the totals. Even as late as 1980, nearly eight in ten Providence residents were white, one in twenty was Hispanic, and only slightly more than one in ten was Black.[65] Within a generation, however, the demographics had changed dramatically. Between 1980 and 2010, when city elders were busily creating the new hourglass economy and the federal government was withdrawing its support for cities, the minority population in Providence nearly tripled.[66]

The Latino population in particular has soared. In 2002, the Pew Hispanic Center named Providence one of the fastest-growing "new destination" cities for arriving Latinos, and in 2010 nearly four of every ten city residents were Latino. The city became majority-minority that year, with Latinos by far the largest minority group in the city.[67] Demographically, no neighborhood has changed more dramatically than Olneyville. In the census totals for 1950, 1960, and

1970, Olneyville was over 99 percent white, and in 1980 it remained almost 97 percent white. Latinos began to arrive in numbers in the 1980s, when the neighborhood had reached its postwar nadir. By 1990, Latinos made up one-third of Olneyville's population, and by 2000 they made up more than one-half.[68]

As in other cities with large Latino populations, their electoral clout has grown with their numbers. The Irish and Italian working-class coalition that had controlled city politics since the 1930s collapsed in the twenty-first century, and in 2010, Providence voters elected Angel Taveras as the city's first Latino mayor. He won with a broad coalition of Latinos, white liberals on the city's East Side, young, college-educated voters, and a large fraction (but not majority) of the Black vote.[69] In 2015, Tavares was replaced by Jorge Elorza, a native son and child of Guatemalan immigrants, who was reelected in 2018 in a landslide. Sabina Matos represents Olneyville on the Providence City Council. Before she took office in 2010, the neighborhood had never been represented by a Latina. In January 2019 she was elected city council president, the first Latina to hold the position.[70]

Several patterns emerge from Olneyville's long history. First, capital is miserly. It has never given more than meager rewards to the neighborhood's workers. For more than 150 years, Olneyville has been a poor neighborhood, barely sustained by work that was monotonous, insecure, and unrewarding.

Second, capital moves. No matter how much of a neighborhood fixture an industry may seem, its position is only as secure as the capital that supports it. And capital always has its gaze fixed on the horizon. It has no allegiance, and when opportunity beckons, it moves elsewhere. Left to its own devices, capital has never been a reliable neighborhood partner.[71]

Third, massive state and federal subsidies brought capital back to Providence in the last decades of the twentieth century, but private development provides no guarantee that all parts of a city will flourish. Capital now supports a bifurcated economy, divided between a large, low-wage, labor-intensive service sector and a small, high-income, knowledge-based information sector. What one scholar observed about the collapse of the garment district in New York applies just as well to the demise of the textile industry in Olneyville: "The tasks of manual labor changed from tending machines to tending people, from sewing garments to changing bedpans and cleaning hotel rooms."[72]

Finally, the "renaissance" that restored Providence was very much a top-down affair. Olneyville, which suffered the most from industrial collapse, had no say in planning the city's future, just as it had no say when the city demolished homes and businesses to build a highway through the neighborhood. Olneyville's fate has always been decided by city elites, the same business and political leaders who excluded the neighborhood from the benefits of the subsidized developments that reshaped the city's downtown.[73] As a result, Providence's revitalization at the end of the century did almost nothing to promote Olneyville's viability.

In short, Olneyville's long history is that of a low-income, working-class, immigrant neighborhood whose residents have repeatedly been let down by tightfisted, footloose capital. At the end of the twentieth century, the political and business classes handed capital enormous subsidies to return to Providence, where it promptly established an economy that was at best indifferent and at worst destructive to Olneyville's longtime residents.

Still, if you are looking for a place to gentrify, Olneyville might be ideal. Providence has bounced back from its postwar slump and now enjoys a reputation as a thriving, vibrant city. Olneyville is close to downtown, with major transportation arteries already in place. It has a long, important history as the center of the textile industry in town, and the Woonasquatucket River wends its way peacefully through the neighborhood. Finally, as a result of long years of neglect, the neighborhood is badly distressed, and land is cheap. Think how valuable it could be, if only the neighborhood could be turned around.

3

"We'll Pray for You"

You've found the neighborhood you want to gentrify. What now? The first challenge might be fixing up the crumbling shells that once provided work for thousands of neighborhood residents. Nothing screams urban poverty like an abandoned factory, with its randomly shattered windows, graffiti-covered walls, and weed-choked parking lot. In Olneyville, these weary sentinels line the streets that run alongside the Woonasquatucket. Empty factories in the neighborhood are surrounded by an eight-foot chain-link fence in a vain attempt to exclude urban refugees seeking an escape from the cost and chaos of the city. A developer who rehabbed a factory along Valley Street in Olneyville told me the construction crew had to drag more than a hundred mattresses out of the building before work could get under way. Walking through one of several abandoned factories in Olneyville, stepping over fencing that had been knocked down by prior visitors, I saw houseplants, kitchen chairs, and art supplies. On a second visit to the site, I spoke with a young man who looked as if he had lived an extremely hard life. His clothes were dirty, he muttered to himself words I could not understand, and as we spoke, his eyes darted around as though he expected trouble to come at him very soon. I asked him if he lived there, and he didn't answer. We didn't talk long.

These buildings can be a magnet for crime and disorder and a drain on nearby property values.[1] They are also an environmental disaster and a fire hazard. A study in Newark found that abandoned buildings are four times more likely than other structures to have a serious fire. Sometimes they burn because homeless people light a fire to stay warm and inadvertently set the building ablaze. More often, however, the cause is arson: a building in a distressed neighborhood might be insured for more than it is worth, and an owner might not be able to afford to maintain it. An investigation in Toledo found that census tracts with the lowest median income had more than fourteen times as many fires of "incendiary" or "suspicious" origin—official code for arson—as tracts with the highest median income.[2] Afterward, instead of an abandoned factory, a distressed neighborhood is saddled with charred wreckage.

How can residents of a distressed neighborhood repair these sites, and what happens when they do?

There is winter, and then there is winter in a New England coastal town. Some places are colder. In others, the wind is more pitiless. Others get even less sun. But when winter sets out to make you miserable, few places can match the New England coast. It was on a particularly bitter night in 1989, one week before Christmas, that the old brick complex wedged between Aleppo Street and the Woonasquatucket River went up in flames. Fueled by oil-soaked floorboards, the fire quickly spread through the buildings that had been Riverside Mills. "Flames 50 to 60 feet high roared out of windows and through the roofs of the complex," United Press International reported. "Sheets of ice formed in the streets as water being poured on the blaze quickly froze in the 10–15 degree temperatures."[3] The fire was so fierce that the Providence Fire Department called in every off-duty firefighter in the city and asked local departments across the state and region to send reinforcements. Firefighters responded to the call from as far away as Foxboro, Massachusetts, a nearly twenty-five-mile drive.[4]

As the fire raged, Olneyville residents who lived nearby fled their homes, fearing that the strong winds would carry embers from the doomed mill to their wood-frame houses. As a precaution, firefighters doused nearby buildings with water, quickly coating them in a protective layer of ice. Michael Moise, the chief of the Providence Fire Department at the time, would later recall the fire as one of the biggest and most dangerous of his long career. "Had we not stopped the fire at Aleppo Street with a water curtain and kept it from crossing the street," he told a journalist after he retired, "with the winds we had that night ... it would have been the destruction of an entire neighborhood."[5] By the time the last fire truck pulled away from Riverside Mills, only one of the original nine buildings was still standing. The Providence Fire Department never established how the fire started, but firefighters and police, as well as several longtime Olneyville residents, suspected arson.[6]

For years, the story ended there. After Riverside burned down, the perennially cash-strapped city simply left the charred carcass to crumble on the ground, a fitting tribute to the collapsed New England textile industry and a familiar back of the hand to Olneyville residents. On the day after the fire, a team from the Rhode Island Department of Environmental Management toured the site and found a chemistry set of toxic waste, including sodium cyanide, acids of various sorts and strengths, waste oil, boiler chemicals, torn bags of asbestos, and barrels of plating sludge. Soon the telltale detritus of urban decay—blown tires and broken appliances—piled on top of the crushed red bricks and twisted, rusting metal. Neighborhood kids frolicked atop the toxic rubble and built makeshift forts from abandoned mattresses, sharing the area with drug users and sex workers.

Mill owners had always relied on New England's rivers, and Riverside, like other mills in Olneyville, had the Woonasquatucket. But for decades, the mills had disgorged tons of raw sewage and chemicals into the waterways. Industrial waste was the largest contributor to the river's pollution. And while the federal Clean Water Act of 1972 led to the decline of toxic industrial waste in the river, nonindustrial pollutants like motor vehicle emissions, household waste, and highway and urban runoff continued to pollute the river.[7] The Woonasquatucket was nearly dead by the time Riverside Mills burned down; no one safely drank the water or ate the few hardy fish that remained. The riverbanks had grown into an impenetrable bramble, and only a few ventured to the waterway concealed beyond. Many people had forgotten it was there.

On the other side of the mill site was Aleppo Street, a narrow residential byway a couple hundred yards long. By the time of the fire, Aleppo Street had long since lost its residential shine, and the blaze only hastened its decline. The lower end of the street opened into Manton Avenue, a busy commercial thoroughfare that runs west through the neighborhood from Olneyville Square; the upper end of Aleppo was a dead end. By the mid-1990s, a squat, sky-blue two-flat was the only occupied house that remained on the street. Two families lived there, one of which actively participated in the neighborhood drug trade. On one occasion, Thomas Deller, then the director of the Providence Department of Planning and Development, traveled to Aleppo Street, where a dealer standing in front of the house pulled a shotgun on him. Past this house, about eighty yards of overgrown weeds and accumulating junk provided ample opportunity for dealers to stash their drugs and guns. Beyond the vacant lot stood the gray concrete back side of a casket factory. Drug users, dealers, and sex workers had discovered a small gap in the space between the new and old foundations of the factory and had dragged in an old couch, which afforded them a bit more privacy and comfort. The police called it the cave.[8]

After the fire, as debris began to accumulate, Aleppo Street became all but impassable. Almost no one came to this swath of land except to buy or sell sex or drugs. Those trades, however, flourished. Dean Isabella, who grew up in Olneyville and had a long career in the Providence Police Department, described Aleppo as "one of the worst, [most] crime-ridden, blighted streets in the city. For decades— and I mean decades—Aleppo Street had been an area used for prostitution, gang activity, drug activity."[9] Robert McMahon, the former superintendent of the Providence Parks Department, once described to me the line of cars that inched down Aleppo Street every day, the drivers picking their way through the debris-filled street to reach the drug dealers in the cave or the sky-blue duplex. The area was known far beyond Providence. Abelardo Hernandez, a longtime resident, community organizer, and devoted advocate for Olneyville, told me that many of the cars in those lines had license plates from outside Rhode Island.

Today the pile of toxic waste that was Riverside Mills is gone, and the site has become Riverside Park. A bike path winds through the greenway, bringing people from the neighborhood to the re-claimed river and connecting the park to a longer path that weaves through the city. Children race around a giant playground—built by Olneyville residents—and clamber over the swing sets and mon-key bars. A well-tended community garden grows there, much loved by local residents. A bicycle cooperative operates under a distinc-tive red tin roof, fixing bikes for local residents and teaching bike repair to neighborhood kids. Local artists have festooned the park with their quirky handiwork. Aleppo Street has been completely re-built. The city cleared away the debris, repaved the street, and opened it on both ends. ONE Neighborhood Builders—the same commu-nity development corporation that administers the Walking School Bus—purchased the one occupied home on the street, relocated the renters, tore down the building, and acquired the vacant lots from the city. Then the corporation built more than sixty units of high-quality housing at below-market rates. Some of the new residents own their homes; some rent. But everywhere the lawns are trimmed, the paint is fresh, the streets are clean, and crime in this corner of Olneyville has all but disappeared.

If Riverside Park has a mother, it is Jane Sherman. Born and raised in West Bend, Wisconsin, forty miles north of Milwaukee, she went to college at Mount Holyoke in western Massachusetts, where she studied economics and sociology. In 1962, during her freshman year, she met a young man who attended nearby Amherst College. The two hit it off, but when Deming Sherman returned to Amherst that evening, he discovered that one of his classmates had also taken an interest in Jane. "We flipped a coin," Deming told me over dinner in the East Side of Providence one evening, "and I won. That was fifty years ago." "He didn't tell me that part of the story for several years," Jane added dryly.

Jane and Deming married after college and moved to the South Side of Chicago. She got a job as a teacher at Florence B. Price Elementary School, at Forty-Third and Drexel Boulevard, while her husband went to law school at the University of Chicago. They lived in an apartment on Fifty-First Street, in the no-man's-land between two legendary Chicago street gangs — the Blackstone Rangers and the East Side Disciples. Jane was at work on Thursday, April 4, 1968, when the news came that Martin Luther King Jr. had been assassinated in Memphis. Like many other American cities, Chicago erupted. The next day, twelve thousand army troopers and six thousand national guardsmen descended on the city. Mayor Richard Daley put half the Chicago Police Department on riot alert. Guardsmen walked the streets two abreast with fixed bayonets.[10] The ghettos on the South Side were spared the worst of the destruction because the Rangers and Disciples temporarily set aside their differences and agreed to keep order, in part as a tribute to King, who had met with both gangs on an earlier trip to the city.[11] Jane still recalls looking out her apartment window and seeing troops patrolling the street below.

After Deming graduated from law school that spring, the couple moved to Providence. Jane got a job teaching at a Providence elementary school but left after a few years to raise her family. Deming joined the storied Rhode Island law firm Edwards & Angell, which brought him into contact with Rhode Island's patrician elite, including Fred Lippitt, whom the *Providence Journal* once described as a "paradigm of Rhode Island's privileged class."[12] A graduate of Yale University and Yale Law School, Lippitt served more than twenty years as a member of the Rhode Island House of Representatives, including a decade as minority leader. His father had represented Rhode Island in the U. S. Senate. His grandfather and one of his uncles had been governor of Rhode Island. His cousin was U. S. senator and former Rhode Island governor John Chafee. Among his many public-service hats, he had been a trustee at Brown University, chairman of the board of Rhode Island Hospital, and a board member of the Rhode Island School of Design, the Nature Conservancy, the Annenberg Institute for School Reform, the Fogarty Foundation for the Mentally Retarded, and the Boys and Girls Clubs of Providence.[13]

For most of his adult life, Lippitt was a Republican, but of a kind that hardly exists anymore. He was part of the eastern establishment wing of the GOP: civic-minded, fiscally conservative, socially liberal, and equally opposed to the southern racism of Jim Crow and the northern corruption of urban machines. Throughout his long career, Lippitt championed racial liberalism and transparent government. But as the center of power in the GOP moved from the Eastern Seaboard to the South and Southwest — from Dwight Eisenhower and Nelson Rockefeller to Barry Goldwater and Ronald Reagan — the GOP's position on race moved sharply to the right. Lippitt, like many others, became increasingly uncomfortable with the party's positions. By the 1980s, Lippitt had left the Republican Party. He ran for mayor of Providence three times, once as a Republican and twice as an independent. His last campaign was in 1990, when he faced off against the legendary Providence machine politician Vincent "Buddy" Cianci.[14]

As I was researching this book, it sometimes seemed that everyone of a certain age in Providence had a Buddy Cianci story. He was thin-skinned, quick to wound, slow to forgive, famously insecure, and desperate for adulation. A politician of the old school, he was at his best in the streets and taverns of Providence, enjoying the cheers and backslapping of his adoring supporters. He used to boast that he would attend the opening of an envelope. But he also ran city hall as his personal cash machine, selling city positions and lucrative contracts like trinkets. The quid pro quo for Cianci's support was participation in his graft. Many who owed their positions to Cianci became his bagmen, soliciting bribes but insulating the mayor from the dirty work.[15] Yet even his harshest critics had to admit that Cianci was not simply a kleptocrat. He was an unapologetic booster for Providence and a tireless champion of the city's revitalization. When I asked people to describe him, more than one person shrugged and said, "Buddy was Buddy." In 2002, Cianci and a number of his aides were convicted in federal court of racketeering and running a criminal enterprise. He was sentenced to five years in prison.

But in 1990, when Lippitt and Cianci faced each other in the mayoral contest, Operation Plunder Dome was still years away. Buddy was on the comeback trail. He had sat in the mayor's chair once before but had pleaded no contest in 1984 to assaulting a man he believed was having an affair with his estranged wife. Forced to resign, he spent the remainder of the 1980s in political purgatory. He became an AM radio talk show host, which kept his name and voice in the public mind, but he was desperate to return to office. By

1990, Cianci had left the Republican Party and ran for mayor a second time. Though he styled himself an independent, he remained a machine politician and campaigned on the slogan "He never stopped caring about Providence."[16]

The race was ugly. Before one televised debate, Cianci sidled up to Lippitt, who never married, and tried to rattle him by taunting him about rumors of his homosexuality. Just before the election, flyers appeared in Black neighborhoods claiming that Lippitt's ancestors in Rhode Island had been slaveholders. Lippitt refused to get into the gutter with Cianci, and voters consistently identified him as the most honest of the candidates.[17] On election night, both Lippitt and Cianci declared victory, the consequence of conflicting projections by the *Providence Journal* (which called the race for Cianci), and the Providence City Canvassing Board (which called it for Lippitt). In the end, Cianci won by 317 votes.[18]

In most places, that would have been the last we heard of Fred Lippitt, who was then approaching his seventy-fourth birthday. If that had happened, Riverside might still be a pile of toxic waste. But Lippitt's civic commitment and Cianci's political acumen combined to produce one more project. In 1992, Cianci asked Lippitt to chair the newly formed Providence Plan, a state-chartered nonprofit created to raise money from private and federal sources for the benefit of the city and state.[19] The plan had only a modest operating budget and could not support organizations or fund programs on its own; its contribution to the well-being of Providence and its residents would depend on the plan's ability to raise external funding.

Throughout its existence, the plan's work tended to reflect the priorities of its chairmen, who served at the pleasure of the mayor. Lippitt, whose family had made its fortune as mill owners, believed the rivers and waterways, which had created such wealth in the state a century before, could anchor its economic and cultural revitalization. He had a vision of industrial brownfields becoming urban greenways. The abandoned factories, junkyards, and vacant lots that lined Rhode Island's waterways could be reclaimed and converted into usable green space that would help transform blighted areas. Many civic leaders, in Rhode Island and throughout New England, shared this vision, but Lippitt was one of the few who carried it to its moral conclusion; he believed the city and state had an obligation to reclaim and restore the segments of the state's rivers that ran through forgotten neighborhoods like Olneyville.[20]

To contemporary ears, accustomed to the take-no-prisoners-make-no-friends style of American politics, Cianci's decision to appoint Lippitt to chair the Providence Plan may seem peculiar. But Cianci was the consummate machine politician and had cultivated the long-lost art of co-opting his enemies by appointing them to lucrative or visible positions. "Marry your enemies and fuck your friends," he used to say.[21] And what better way to fuck Fred Lippitt than to hand him the herculean challenge of cleaning up toxic rivers like the Woonasquatucket, with no budget to do it? If Lippitt failed, he would get the blame. If by some miracle he succeeded, Providence would benefit, and Cianci would take the credit.

In late 1993, Lippitt asked Jane Sherman to join him at the Providence Plan and charged her with developing a greenway along the portion of the Woonasquatucket that ran through Olneyville. Sherman, like Lippitt, had a long history of service. She had volunteered for dozens of community and civic organizations, from the Rhode Island Rivers Council to the League of Women Voters. In our many meetings, Sherman was modest about her public service. I only discovered it by tracking down a bio online, which was itself part of an announcement marking her appointment in 2011 to the Rhode Island Board of Governors for Higher Education.[22] When Providence celebrated its 350th anniversary, Sherman was honored as one of the city's outstanding community leaders, and that was *before* her work in Olneyville. When I asked her about all of this, she shrugged. "When I was younger," she said, "I did a lot of better-government stuff." She also had a connection to Cianci, who had appointed her to the Providence City Planning Commission, another volunteer position, during his first stretch as mayor. She brushed that off as well. "I had been the president of the League of Women Voters, and he wanted a good-government person." Marry your enemies.

Sherman's work in Olneyville could easily have gone sideways. A well-off white woman, with a long history of volunteer service and a comfortable life on the moneyed East Side, comes down to one of the most distressed, impoverished neighborhoods in the entire state. Her husband's law partners include some of Rhode Island's most prominent public figures. Now she plans to "fix" a low-income Latino neighborhood in the throes of the crack epidemic—a neighborhood that has been decimated by decades of decline and capital flight and ignored for years by Rhode Island's power brokers. To top it off, she has no budget. Mindful of the optics, Sherman was careful not to make promises she knew she couldn't keep. From the moment she set foot in Olneyville, she let everyone know she had no resources. "I was really very frank," she told me. "The mayor said, 'Try whatever you want, but there's no money.'"

Sherman brought groups from all over Providence to visit the pile of twisted junk along the banks of a nearly dead river. "The Department of Environmental Management. The Southern New England chapter of the Soil and Water Conservation Society. The Army Corps of Engineers. Anybody who'd go. Just trying to drum up support" for what surely seemed a cockamamie plan. Sometimes her guests would bundle into a tour bus and ride over from their offices downtown, but mostly she shuttled them around the neighborhood in her car.

Her visitors were skeptical. "Where are you *taking* me?" a woman asked on one occasion. They would step slowly out of the car and look around uneasily at the abandoned buildings, accumulated rubble, and knots of drug users and sex workers. Some refused to get out of the car. The intrepid ones would pick their way over crumbled bricks and step gingerly around discarded needles as Sherman laid out her vision for the site: "All of this will be a greenway along the river," she would say with a sweep of her hand, "with a bike path that winds along here and connects Olneyville to the city. We'll put a playground over there, and maybe a community garden just beyond." But all anyone could see was a pile of junk. After one tour, as the group settled back into the bus, she asked, "Well, what do you think?" There was a pause, and then a man called out from the back, "We'll pray for you." She chuckled with the others but thought ruefully that prayers would probably not get the job done.

People in the neighborhood were no less skeptical. At one point, Sherman wanted to hold a festival in Donigian Park, on Valley Street along the Woonasquatucket, to showcase the river as a neighborhood asset. But she decided against it after concluding that too many syringes littered the ground. It's not for nothing that local residents used to call Donigian "Needle Park." "It just wasn't safe," she told me matter-of-factly. They held the festival in a parking lot. When Sherman realized they needed an electrical outlet for their sound system, she asked a local business owner to leave a window in his shop open over the weekend so that she could plug it in. "He was like, 'Sure, I'll leave the window open,'" shaking his head at the lunacy of Sherman's ambition, as though she had asked him to fund her personal moon launch. "People were supportive, but skeptical. They just didn't think anything was going to happen. And that was okay. I was skeptical too. We put up a tent and brought in canoes and just talked about the possibilities. People came by and asked, 'Where did you get the river?' They didn't know there was a river there."

A public park is communal space, and it works only if the neighborhood accepts it. If it is rejected by the residents who live nearby — if they do not use it, care for it, and insist that the city do the same; if they do not take ownership over the park's design, creation, and life — it can fall into disuse. At the extreme, it can degenerate into a magnet for crime and disorder.[23] Precisely because this relationship is so well known, many funders will not contribute money to a reclamation project unless they are convinced the plan genuinely represents the will of the community. Acceptance by the local community is not the only condition of a successful urban park, but it is certainly a necessary one.[24]

Every path looks easy once it is cleared, and the success of Riverside Park can create the misleading impression that engaging the residents of Olneyville was a simple matter. But it is not easy getting the men and women in a badly beaten neighborhood to believe that city elites would suddenly go to bat for a place they had long ignored. Consider the situation from the perspective of an Olneyville resident. The city had funneled billions of federal dollars into the downtown but had allowed Riverside Mills to fester for years as a toxic waste site. Aleppo Street had been a haven for criminal activity for decades. Crime rates, drug use, and unemployment were among the highest in the city. When Sherman started her work in Olneyville, the neighborhood had zero functioning parks: Donigian had been all but taken over by drug users; Joslin Park, which abutted the local elementary school, was so dangerous that the school did not send kids outside during recess; and Merino Park, in an adjoining neighborhood, had been closed by the city because of its isolation and security issues.[25]

Is there anything in this profile that would have led people in Olney-ville to believe the city cared one whit about what they wanted? Or that the city would make good on a vague promise to build a new park when it had allocated no money for the task and couldn't even maintain the parks it already had?

Sherman told me she "walked the streets of Olneyville for about a year, listening." She talked with any neighborhood group that would meet with her, and organized countless meetings in community centers, church basements, half-empty restaurants, and private homes, trying to engage the neighborhood and figure out what residents wanted from the Riverside Mills site. By this time, she had also received a modest grant from a private foundation that allowed her to hire an assistant. Lisa Aurecchia is a lifelong Rhode Island resident. Her Jewish maternal grandmother had escaped Austria in 1939 when Lisa's mother was an infant. The family moved originally to New York but soon relocated to Providence, where they settled in Olneyville, living on Delaine Street, a few blocks from Riverside Mills. As Jane turned to fundraising and drumming up support from outside Olney-ville, the community mobilizing passed to Lisa.

Where Jane brings a steely dignity to her work, Lisa has a profanity-laced directness. Jane comfortably mingles with senators; Lisa hates the limelight. It took a long time before she agreed to an interview, and when we finally talked, it quickly became clear she is more interested in the next challenge than the last. Her recollections spit out like water from a faucet that hadn't been turned on for years. I asked her how they got the word out. "We just went. . . . We were going to every community center, rec center, after-school program. . . . There were three major churches . . . Olneyville Public Housing. We had a slideshow. . . . We went to every classroom." Sounds like you were focusing on young people, I observed. "We hired kids. We called them River Rangers. Cleaning up, painting over graffiti. Paid them a little." What about adults? "Oh yeah, it wasn't just kids. We were flyering. We went door to door. We flyered the *fuck* out of the neighborhood." How did you figure out what people wanted? "Surveys. We did hundreds of surveys. I think I just threw those out." I mentioned the festival they held in a parking lot rather than Donigian Park. "Donigian Park was a shit show," she said. Lisa has now been working for Olneyville for more than twenty years. She feigns a war-weary cynicism but is one of the most committed, passionate advocates I met in my research for this book.[26]

In 1997, after more than three years of public meetings, private buttonholing, community festivals, and anticrime marches, Sherman and Aurecchia finally had a master plan that reflected the neighborhood's input. The plan called for the revitalization of the Woonasquatucket, the remediation of the mill site, and the construction of a community park, complete with a bike path. But who would fund the creation of a park in a place like Olneyville? Was the plan destined to gather dust on a shelf, like so many other grand but unfunded designs?

In his State of the Union Address in January 1997, President Bill Clinton announced the American Heritage River Initiative, a program designed to draw attention and resources to rivers in the United States that had made the greatest contribution to the nation's economic and cultural life. Four months later, Vice President Al Gore announced the creation of the Brownfields National Partnership. The partnership promised to bring federal resources to support local remediation and revitalization of abandoned and contaminated industrial properties known as brownfields.[27] Sherman immediately understood the potential of the two initiatives. If she could get the Woonasquatucket designated as an American Heritage River and Olneyville designated as a Brownfields Showcase Community, it would bring resources and attention to her project. But what were the odds? The United States has thousands of rivers, and the EPA estimates there are more than 450,000 brownfields. Of these, the Clinton administration initially planned to designate just ten American Heritage Rivers and create a partnership with sixteen Brownfields Showcase Communities.[28]

In 1997, Sherman reached out to the staff of John Chafee, Rhode Island's senior senator. Chafee, like his cousin Fred Lippitt, was a fiscal conservative and social liberal who served Rhode Island as a Rockefeller Republican for years. He was pro-choice, supported gun control and the ban on school prayer, and opposed the death penalty. But his greatest and most enduring achievements may have been for the environment. As chair of the Committee on Environment and Public Works, he played a prominent role in the 1990 reauthorization and expansion of the Clean Air Act, directing significant federal resources to the fight against acid rain.[29]

Chafee incurred the lasting ire of some conservative members of his party by using the chair's prerogative to block consideration of bills that would have dismantled existing environmental protections. When Chafee died in 1999, Carl Pope, the executive director of the Sierra Club, described him as a "one-man front line" against the "far-right wing" of the GOP. "No one will ever know how many bad things did not happen in the last three years because John Chafee was there."[30] Chafee's commitment to the environment would prove pivotal to Riverside Park. When Sherman met with his staff, she explained the plan for the Woonasquatucket and asked for the senator's help. Would Chafee nominate the river for designation as an American Heritage River? He would.

At the same time, Sherman also reached out to Jack Reed, Rhode Island's junior senator. Reed, who had earlier been a lawyer with Deming Sherman at Edwards & Angell, had served three terms in the state Senate followed by another three in the U.S. House of Representatives. He has represented Rhode Island in the Senate since 1996. Like Chafee, Reed has an abiding commitment to the environment and quickly recognized the value of Sherman's project. He even joined Sherman on a tour of the site in the mid-1990s, trudging through knee-deep snow. Early in 1998, Reed nominated Olneyville as a Brownfields Showcase Community.[31]

In March of that year, Vice President Gore approved Reed's request and named Olneyville a Brownfields Showcase Community, with support intended for the remediation of two sites along the Woonasquatucket, including Riverside Mills. Four months later, President Clinton designated the Woonasquatucket an American Heritage River.[32] The tiny Woonasquatucket was one of only two rivers in the country to win both designations. Though they did not guarantee the ultimate success of Riverside Park, they certainly were instrumental in shaking money loose. In 1998, the EPA awarded a grant to conduct a remediation plan for the site, and in the following year, the State of Rhode Island agreed to contribute slightly less than half a million dollars toward the cleanup. In 2000, Senator Reed secured a $1 million grant from the Department of Housing to pay for the remediation.[33] In the same year, and more than a decade after the fire that destroyed the mill, Providence finally made a financial commitment, promising another $1 million in bond funds.[34] In the late spring of 2000, remediation finally began.

Removing the surface debris was slow but straightforward. So was removing the poisonous standing water, which was pumped into a pair of 20,000-gallon steel tanks. Regrading the surface and building two retaining walls to prevent erosion into the river likewise presented no unusual challenges. But no remediation process ever goes according to plan. "You never know what you're going to find when you start to dig," Thomas Deller of the Department of Planning and Development told me one afternoon. The remediation crew, for instance, uncovered two long-forgotten underground storage tanks. The smaller of the two contained about 375 gallons of fuel oil; the larger was a 20,000-gallon tank that had been emptied of most of its oil, filled with soil, and abandoned while the mill was still operating.[35]

Then there was the problem of uncounted tons of polluted soil. Some of the land was so toxic that contractors had to haul it off in huge dumpsters. The rest had to be covered in two feet of fresh soil. And because Mother Nature always gets the last word, a once-in-a-century flood in October 2005 washed away nearly all the fresh soil only days before workers were scheduled to lay sod on the site, requiring a new layer of soil. Finally, before the contractors could dump clean dirt onto the poisoned land, before they could regrade it or build a retaining wall, they had to lay a specially made polypropylene fabric over the entire site ("a non-woven, needle-punched, permeable geotextile fabric ... to prevent human exposure to impacted soil"), a space-age cloth that would permanently entomb the toxic land. It was an ironic but fitting end for a mill that had started life making woolen coffin covers for Union soldiers killed in the Civil War.[36]

All told, the remediation of Riverside Mills and construction of Riverside Park cost more than $2 million and took eight years.[37] Is a small urban park worth all this time, money, and effort? Academics and nonprofit organizations have tried to answer such questions by assigning value to what cannot be readily measured. The Trust for Public Lands, for instance, has devised an elaborate formula to estimate the economic benefit of a well-designed urban park, including, among other factors, the increase in adjacent property values and the benefit from rising wealth that washes through a neighborhood because of its park system. (Studies show that the best parks—of which Riverside is clearly an example—add approximately 15 percent to the value of adjacent property.) Others, like the National Recreation and Park Association and the City Parks Alliance, have calculated the contribution of city parks to public health, stamping a number on the forehead of the people who don't have a stroke, who see their blood pressure fall and their diabetes successfully managed. Studies have tallied the savings from greenhouse gases that are not pumped into the air, the temperature that does not rise, and the soil that does not erode because of a park's green footprint.[38]

This research has great value, if only because many people in this neoliberal moment believe that every expenditure of public funds demands a prior showing that the benefit of going forward far exceeds the cost of inaction. For those who think in such terms, the answer to the question "Is it worth it?" is a resounding yes. The economic, cultural, environmental, public-health, and educational value of a well-designed city park far exceeds the cost of creating and maintaining it.[39] But I have always believed these arguments miss something important. No one demanded such a showing decades earlier when Olneyville was wrecked in the name of urban renewal, when city planners carved a highway through the neighborhood, destroyed thriving businesses, or tore down working-class homes. No one undertook to calculate the cost to Olneyville residents of leaving a toxic waste dump in their backyard for more than a decade. The neoliberal acolytes who are so quick to demand fiscal restraint when we want to heal an urban neighborhood were not so punctilious when the neighborhood was first made sick.

If we were to mark the beginning of Olneyville's revitalization, it would be when the charred wreckage of Riverside Mills was transformed into one of the best parks in Providence. But I have not dwelled at length on the creation of Riverside Park simply because of its importance in Olneyville's story. It also tells us a great deal about the promise and limits of the current approach to neighborhood well-being in a neoliberal age. Riverside required the civic-minded volunteerism of Fred Lippitt; the municipal entrepreneurism of the Providence Plan; a start-up contribution from private philanthropy; support at a critical juncture from senior elected officials; scientific expertise and funding from the state, local, and federal governments; and the mobilization of local residents, all shepherded by Jane Sherman, an indefatigable leader with a long commitment to progressive social change at the local level. The park shows the great potential of the current approach, at least when the pieces line up just right. It transformed a toxic dump into an enduring asset that dramatically improved the lives of Olneyville residents.[40]

Yet the park also shows the limits of this approach. On one occasion, Jane Sherman and I met at a deli on the East Side of Providence, near her home. The coffee was awful, but the quiet setting was perfect for a long interview. In keeping with her commitment to behind-the-scenes community service, Sherman is reluctant to take credit for anything. On this occasion, she insisted that the success at Riverside Park was just luck. "All the stars aligned," she said with a shrug. "Clinton, Chafee, Reed. The American Heritage designation. The Brownfield designation." She could have added the one degree of separation that brought her in touch with Fred Lippitt, their mutually quirky connections to Buddy Cianci, or Lippitt's familial link to Chafee and professional link to Reed. "It could never happen today," she said.

In one sense, she is certainly wrong. There is nothing unique about Riverside Park, from its history as a toxic waste dump to its transformation into an urban oasis. The expertise needed to make the change has been widely available for decades. In fact, places like Riverside Mills have also been reclaimed in other cities.[41] It *could* happen today, in Providence or anywhere else in the country. Yet in an age of neoliberal precarity, Sherman is more right than wrong. What are the odds that an activist at another city's industrial waste dump will be able to line up the dominoes as they lined up here? But what Sherman attributes to luck, I would credit to wealth and power.

Without taking anything away from Sherman, I contend that Riverside Park worked because elite political power at the local and national levels unleashed massive infusions of capital. If power and money had not joined in this way, Riverside Mills might still be a toxic junkyard. What kind of approach to community revitalization relies on a tangled, idiosyncratic web of political, professional, and familial connections that can only be replicated if financial, political, and bureaucratic forces align in a way that is almost impossible in a poor neighborhood? What of the distressed Providence neighborhoods that do not happen to sit next to a polluted waterway and thus do not have champions like Jane Sherman and Fred Lippitt? What of the distressed neighborhoods all over the country, whose advocates cannot claim a personal connection not only to the mayor but to *both* U.S. senators, including a powerful committee chair? The approach that worked in Riverside tends to succeed only when it is in the interest of the wealthy and powerful to allow it to work, which once again makes the neighborhood prey to shifting political and financial priorities set by outsiders.

And in a neoliberal age, those priorities have indeed shifted, away from government and toward the creation of private wealth. Since the mid-1990s, federal responsibility for brownfield remediation has fallen to the Environmental Protection Agency, which as of May 2020 had awarded nearly $1.6 billion in grants to assess and clean sites.[42] This might sound like a lot of money, but of the more than 450,000 brownfields in the nation, the EPA had made fewer than 7,800 ready for reuse by October 2019.[43] Measured against the magnitude of the need, the federal commitment is trivial, and the EPA candidly admits the brownfields program is "not intended to address all of the brownfield sites in the U.S."[44] The agency says it denies two of every three requests for funding.[45] Having a Jane Sherman helps.

And who gets the small amount of money doled out? The EPA anticipates that the heavy lifting will be done at the local level, by either local governments, the private sector, or a public-private partnership.[46] For most cities, the first option is laughable. Robert McMahon, Providence's deputy superintendent for parks during the Riverside project, told me the average budget to repair or reclaim a park during his tenure was $50,000. Riverside cost forty times that. Thus, as a practical matter, the EPA expects that brownfield remediation will be handled by the private sector, either on its own or in partnership with state and local government. For-profit developers, however, have no interest in cleaning up a toxic dump just to build a park. They want to create something that will make money for investors. The EPA is perfectly explicit on this score: it acknowledges that the interest of real estate developers in a brownfield project is to "maximize return on investment and complete the deal in the shortest amount of time possible," while the interest of potential buyers at the site is to "maximize profits through reselling the property or long-term ownership and management."[47] If developers build a park, therefore, it is apt to be an amenity attached to a profit-making development like a shopping district or upscale housing.[48]

Because brownfield redevelopment focuses on making money, remediation frequently (but not always)[49] takes place in comparatively wealthy areas, where development risks are low, or in areas on the cusp of gentrification, where the cost of land is low but developers expect it to rise dramatically, perhaps as a result of the remediation. In Massachusetts, for instance, most cleanup funds have been funneled into suburban areas, which are more attractive to developers, rather than distressed urban areas. Coralville, Iowa, has a paid staff member responsible for coordinating its $40 million River Landing District, which is set to include townhouses, a theater, and hotels at the site of a former truck stop, scrapyard, and warehouses. Coralville, a middle-class neighborhood of nineteen thousand with 14 percent living below the poverty line, has received $1.9 million in EPA grants since 1999, the most money given to any city in the state.[50]

The EPA makes no secret of its preference for wealthier areas: it advises developers to "evaluate whether development opportunity outweighs the risk." The agency identifies key opportunities for developers when taking on a brownfield remediation, including securing land in a "prime location," thus affording a chance "to acquire and develop a property in a highly desirable location where there may be few or no other properties available," and acquiring contaminated property "at a reduced cost, which could potentially increase profit."[51] By contrast, the EPA cautions residents of distressed neighborhoods (which it tellingly describes not in human terms but as areas with "a weak market") that their "desire for redevelopment" may not be matched by the presence of a willing developer, and they may have to explore "alternative" redevelopment strategies.[52]

To sweeten the pot for developers, local governments typically offer incentives that lower development costs and increase potential profits. A municipality, for instance, might acquire the land by condemnation or eminent domain and sell it to developers at a reduced price or build necessary infrastructure like roads. Once again, the EPA considers such efforts part of the brownfield redevelopment "anatomy."[53] Municipalities likewise consider these incentives essential; as the former brownfields coordinator for the State of Connecticut once observed, remediation projects "are real estate transactions and real estate projects and if the development has no likelihood of success, that process will likely not result in a cleanup."[54]

The neoliberal approach to brownfields has resulted in a phenomenon that scholars call environmental or green gentrification, which refers to the link between gentrification and brownfield remediation. A researcher with the EPA, for instance, examined demographic changes at sixty-one brownfield sites and concluded that "gentrification is often a consequence of brownfields redevelopment."[55] The causal arrow is complex: sometimes remediation leads to gentrification; sometimes gentrification leads to the remediation of contaminated sites that had been neglected for years. Yet central to each scenario is the morally obscene paradox at the heart of this book: in an age of neoliberal precarity, improvements in a poor neighborhood may destroy what the poor have labored so hard to achieve.[56]

Finally, even if brownfield remediation were targeted to places like Olneyville, and even if it did not lead to gentrification, the long journey that led to Riverside Park had yet another consequence. Think about the many connections forged and the skills developed in the process of creating the park. When Sherman and her team finished, Olneyville had a magnificent park. But the residents of Olneyville had no connection to Senators Chafee or Reed that they could leverage for future projects. Those connections resided with Sherman and her team. When Sherman and Aurecchia successfully mobilized the residents behind the park, they tapped the enormous potential of an engaged neighborhood, but their organizational skills did not accrue to the residents themselves. Those skills remained with others.

In 2001, Sherman peeled off from the Providence Plan to create the Woonasquatucket River Watershed Council, which today is by far the most important organization in the state in reclaiming and protecting the entire Woonasquatucket corridor as usable greenspace. The council remains active in Olneyville and, among other achievements, has extended the bike path that began at Riverside Park throughout the neighborhood. Yet it still relies on the organizational tools and structure developed more than two decades ago by Jane Sherman. Though the Watershed Council unquestionably benefits

Olneyville residents, it remains an outside organization and does not provide expertise, training, or organizational capacity that neighborhood residents can use to control their own destiny.

In a world of neoliberal precarity, thinly funded nonprofits can repair infrastructure that has been badly damaged by years of environmental pollution and neglect—if all the stars align. But as Jane Sherman understood, success often depends on marrying wealth to power and putting them to work for the poor rather than the wealthy, which turns out to be every bit as improbable as it sounds. Indeed, it requires resisting the strong forces created by government policy that explicitly favors developers who are foremost interested in making a profit. And even when this approach works, the social capital, organizational benefits, and institutional expertise created by the effort flow principally to nonprofits rather than neighborhood residents. As a model of neighborhood well-being, therefore, the current approach can (sometimes) make the physical space vastly more attractive, but it cannot empower neighborhood residents to resist the consequences of that attractiveness.

4

"We'll Be Happy to Indulge Your Fantasies"

You've found the neighborhood you want to gentrify, and getting the residents and their allies to clean up that toxic junkyard was certainly a big step forward. But that's only the beginning. What about all those abandoned houses and vacant lots? Like the junkyard, they depress property values and attract crime. If you want to gentrify the neighborhood and displace its residents, the next step might be fixing the housing stock.

Not everyone in Olneyville thought Jane Sherman was crazy. As a child growing up in the working-class Boston neighborhood of Jamaica Plain, Frank Shea heard stories about the highway that never was. In the 1960s, city planners were poised to rip through the heart of Jamaica Plain to make way for eight lanes of Interstate 95. Like many other cities, Boston wanted to create an elevated highway so that people in the fast-growing suburbs could easily drive downtown. Hundreds of shop owners and thousands of residents would have been ripped out and tossed aside like weeds with shallow roots. If the experience in Olneyville and other parts of the country was any indication, the businesses would never have reopened, and the neighbors would never have returned. And the decline that was already beginning in Jamaica Plain would have accelerated.[1]

But something extraordinary happened. The residents of Jamaica Plain and other neighborhoods rebelled and blocked the highway, forcing the city to cancel its plans for I-95. Instead of an above-ground highway, the city built a below-ground commuter line—the Massachusetts Bay Transportation Authority Orange Line. On the surface, the city built Southwest Corridor Park, an almost five-mile-long ribbon of green from Back Bay Station in Boston to Forest Hills Station in Jamaica Plain. The award-winning park is a masterpiece of urban design, and the grassroots activism that blocked the highway and brought the park into existence is a much-admired case study in citizen resistance to government hubris disguised as urban planning.[2]

Yet Southwest Corridor Park was shortsighted in one critical respect. When construction started in the late 1970s, no one thought Boston and its surrounding suburbs would grow as they have over the last four decades or be reborn as a center of the tech and information industries. In 1980, Boston's population had fallen to a little over 560,000, its lowest level of the twentieth century. By 2018 the city had added about 130,000 residents, with its growth concentrated in the high-paying sectors of the information economy that have proved irresistible to young, well-educated, well-paid newcomers.[3] Southwest Corridor Park and the commuter line beneath it put Jamaica Plain in a perfect position to benefit from Boston's explosive growth. And the one thing no one had planned for was affordable housing. The park that saved Jamaica Plain also laid the groundwork for its eventual undoing as a vibrant, affordable working-class neighborhood. Shea grew up in a modest three-flat in Jamaica Plain that his parents bought in 1962 for $12,000. In 2005 and '06, he and his siblings sold the three apartments for roughly $300,000 *per unit*. A dozen years later, in a similar building across the street, a single unit sold for $535,000.

"What happened in Jamaica Plain was really formative for me," Frank told me the first time we spoke. We met for breakfast at the Classic Café, a diner on Westminster Avenue in Providence, a few miles east of Olneyville. Before he graduated from Harvard in 1984,

Shea volunteered for Boston city councilor Ray Flynn in his successful campaign for mayor. After Flynn won, Shea joined the new administration, eventually becoming part of the Office for Neighborhood Services, which Flynn established to make good on his campaign promise to revitalize the role of local neighborhoods in city planning. In 1993, the year Frank and his wife moved to Providence for his wife's medical residency, he began working for the Metro Boston Housing Partnership, where he acquired an expertise in affordable housing.

In January 2000, a few months before remediation began at the Riverside Mills site, Shea became executive director of the Olneyville Housing Corporation, a nonprofit community development corporation formed in 1988 and dedicated to improving the quality of life in Olneyville. Today the organization is called ONE Neighborhood Builders, which is how I refer to it. When Shea started there, the neighborhood was suffering. "Not a lot of investment," he told me between sips of coffee. "Not a lot of forward thinking." He repeated a trope that I heard several times during my research, always attributed to a former senior official in the Cianci administration: the best thing Providence could do for Olneyville was to plow it into the river and create an industrial park. As for Jane Sherman's seemingly harebrained plans for Riverside, Shea knew the conventional wisdom. "No one thought it would happen." Even the neighborhood residents were skeptical. "They were like, 'We'll be happy to indulge your fantasies, but ...'"

Shea saw things differently. When he heard Sherman's plans for the park, he thought about Jamaica Plain. "It was obvious," he said, "number one, that they could *do* it, and number two, that it could be transformative." Yet he also understood that the enormous investment in the park and all that Sherman and her team had done would be for naught unless someone also reclaimed Aleppo Street. Few would use the park if getting there meant running a gauntlet of drug dealers, sex workers, and gang members. That kind of half measure had doomed the remediation of other urban parks around the country.[4] Perhaps in time the private sector would be given enough government subsidies to rebuild Aleppo Street, but there was certainly no guarantee. And even with government funding, rebuilding could take years, and when it finally happened, the houses would be priced well beyond the reach of the people who lived in the neighborhood, as had happened in Jamaica Plain. For Riverside Park to work for the residents of Olneyville, Aleppo Street had to be rebuilt as affordable housing.

We overuse the word *crisis* in this country, and many conditions described as crises are not. But the greatest challenge facing any city today is to provide safe, healthy, affordable shelter for its residents. Good housing is much more than simply a roof and walls, and we begin to understand its importance when we consider what happens in its absence. There is, for instance, the simple challenge of staying healthy, since a person who lives in substandard housing is more apt to get sick. Poor housing is associated with higher rates of asthma, lead and other environmental poisoning, and elevated stress and assorted mental health problems.[5] Then you have the contribution that quality housing makes to a child's education. Children in safe, stable homes tend to get more out of school. The research suggests that high-quality, affordable housing can promote the educational success of low-income children "by supporting family financial stability, reducing mobility, providing safe, nurturing living environments, and providing a platform for community development."[6] The same applies to public safety: public housing units financed by low-income housing tax credits can lead to significantly lower crime rates. This should not be surprising, since this housing frequently replaces dilapidated buildings and abandoned lots that are a magnet for crime.

Combine this with the regulations that require that these units be well maintained, and you predictably increase the likelihood of a safe, vibrant neighborhood.[7]

Because quality housing is tied to so many other positive outcomes, most people will make great sacrifices to get the best housing they can afford. As housing becomes more expensive, they skimp elsewhere. Many people end up spending an extraordinarily large fraction of their income on housing. The federal government reckons that a family should pay no more than 30 percent of its income on housing; more is not considered financially prudent. Yet poor families, including the working poor, *routinely* spend much more than this on shelter. More than half of all poor renting families spend over 50 percent of their income on housing, and one in four spends more than 70 percent.[8] When families are forced to pay so much to keep a roof over their heads, they often forgo things that wealthier people consider essential. One study, for instance, showed that working families who paid 50 percent of their income on housing only spent half as much on health care and insurance as families living in affordable housing.[9]

Given the singular importance of quality, affordable housing, one might think the United States would make building such homes a national priority, and politicians would invoke the chasm between what is and what ought to be as a measure of the distance we must travel before we dare claim to be a just society. And that distance is great indeed. The conventional measure of the housing crisis combines the number of households that spend more than 30 percent of their income on shelter—they are considered "burdened"—with the number of households that spend more than half their income: the "severely burdened." The total is staggering. As of 2017, 20.5 million renting households and 17.3 million home-owning households were burdened, or nearly a third of all households nationwide. Another 18 million households are severely burdened, the great majority of which survive on less than $30,000 per year. On average, severely cost-burdened families with children spent only $310 per month on food, about half the cost of the most minimal food plan the U.S. Department of Agriculture recommends for families.[10]

As many scholars have pointed out, however, these categories have to be refined, since they do not account for differences in expenses and income. The rich can spend substantially more than 30 percent of their income on housing and still have plenty left over to live a healthy life in a safe neighborhood. A single person living alone might be able to spend more than 30 percent on housing if his or her other expenses are low. A large family with substantial expenses for medical care, clothing, and transportation, however, might find that spending 30 percent of household income on housing is simply unmanageable. The better question, therefore, is not what fraction of the household income a family spends on shelter but what they must sacrifice to pay the rent or mortgage. As the University of Massachusetts professor Michael Stone puts it, how many people in the United States are "shelter-poor"? Stone estimates the total is about 90 million, but his approach allows us to zero in on the people who genuinely cannot afford their housing. Predictably, Stone found that families headed by a single breadwinner, especially within communities of color, where incomes tend to be low and outside expenses high, are most apt to be shelter-poor. This profile is common in Olneyville, where nearly 60 percent of the children—almost twice the national average—live in households with a single parent. Nearly half of those households survive on incomes below the poverty line.[11]

Why are millions of people in the United States unable to find affordable housing? Blaming the poor for their plight has long been a popular parlor game. But a problem that reaches more than one of every four Americans cannot be dismissed as a failure of personal responsibility. It is the result of a half century of structural changes in the American economy—transformations entirely beyond the control of individual workers—that were amplified and reinforced by law and policy at all levels of government. The poor did not slouch their way to the world they now occupy; they were forced into it by a toxic combination of economic reality and political preference.

To begin with, wages have flatlined. After controlling for inflation, wages were only 10 percent higher in 2017 than they were in 1973, representing an annual real wage growth of below 0.2 percent. Economists disagree about why wage growth has disappeared. Their explanations range from the decline of unions and the rise of noncompete clauses to a lagging minimum wage and increasing globalization and automation.[12] But no one credibly challenges this basic, sobering fact: workers are paid basically the same today as they were 40 years ago.[13] Yet according to the Economic Policy Institute, more than half the states (including Rhode Island) have passed laws prohibiting cities and counties from setting a minimum wage higher than the state minimum, even when local costs are substantially higher than the state average. Nearly half have also passed laws that prevent cities and counties from requiring local employers to offer paid sick leave or other forms of family or medical leave (again including Rhode Island).[14]

Meanwhile, the cost of shelter has skyrocketed. Between 1970 and 2000, the average cost of a home in the United States, after controlling for inflation, nearly doubled. The cost of rent has also consistently outpaced inflation. Between 2000 and 2015, the median rent in the United States, adjusted for inflation, increased by about 10 percent.[15] Because this nationwide calculation includes many places in rural America where populations and rents are falling, it tends to understate the severity of the problem in cities where rents are rising. In Providence, for instance, the median rent over the same period increased 26 percent. And perversely, rents are rising fastest—far faster than inflation—for those in the lowest income brackets. In Olneyville, the median rent increased more than 50 percent between 2000 and 2015.[16] In other words, though all renters in the city are spending proportionally more on shelter, rents are increasing the fastest for the people least able to manage the cost.

Adding to the strain, the ranks of the working poor have exploded and remain high, even after years of economic expansion. In 2019, a paper by the Brookings Institution reported that more than 53 million people, or 44 percent of all workers aged eighteen to sixty-four, now work low-paying, hourly wage jobs.[17] For many of these workers, financial security is a contradiction in terms. According to a 2015 study by the Pew Research Center, one-fifth of all adults in the United States, or just under 50 million people, live in a household that is in or near poverty. In 2015, that meant they lived on less than $31,402 per year.[18] But even that income lies far beyond the reach of most households in Olneyville. In 2015, the median household income in the neighborhood was a little more than $24,000. Three years later, it had climbed to just over $27,300. In 2018, after eight straight years of economic growth, more than a third of Olneyville households were living on less than $15,000 a year.[19]

Finally, work itself has become more precarious, even for people with a college degree. As the service sector has eclipsed manufacturing, the United States has seen an increase in what some scholars call "bad jobs." Workers are less likely to have job security and more likely to face extended periods of unemployment. Even when they have jobs, they are less likely to have benefits such as health insurance, sick leave, or childcare. Though every sector of the economy has felt the impact of this shift from employment security to insecurity, "bad jobs" are held disproportionately by the working poor.[20]

Add these structural trends together—no wage growth, higher

housing costs, and less job security—and you have a housing crisis. It has been with us for years. In 1996, the journalist Jason DeParle observed in the *New York Times Magazine* that "the low-wage jobs of the new economy cannot pay the rent." The high cost of housing, he wrote, either "breaks the budgets of low-income Americans or crowds them into violent ghettos, far from good jobs and schools— or both."[21] Since DeParle wrote these lines, things have gotten much worse. Articles now appear almost daily decrying the nation's persistent failure to provide affordable housing for the working poor, leaving an appalling fraction of the population on the knife-edge between housed and homeless. As I was writing this chapter, Matthew Desmond, the Princeton sociologist and author of the Pulitzer Prize-winning *Evicted: Poverty and Profit in an American City,* released his expanded database on evictions in the United States. Among his findings: 900,000 families were evicted nationwide in 2016, or roughly 2.3 million people made homeless in a single year.[22]

Contrary to what some want to believe, the free market will not solve this crisis. The cost of acquiring land and building homes in most urban markets is simply too great. Unless private developers receive substantial ongoing subsidies, they cannot and will not build and maintain affordable housing. In many of the tightest markets, where construction and land costs are exceedingly high, developers demand huge subsidies even for market-rate or luxury housing. Until recently, a state law in New York provided developers with a substantial tax break if at least 20 percent of the apartments they built were set aside as affordable housing. But a provision in the law allowed luxury developers to satisfy this requirement if they made small investments in affordable housing elsewhere, far from their luxury apartments. This led to some eye-popping absurdities. At One57, for instance, the über-luxury apartments that rise above Fifty-Seventh Street in Manhattan, apartments have sold for as much as $100 million. Yet the developers relied on the tax abatement to cut their property taxes by 95 percent. In exchange, they built sixty-six affordable apartments in the Bronx. When the abatement program came under attack, developers warned that without the subsidies, building even luxury apartments in New York would be nearly impossible given the costs of land and construction.[23] If the private sector cannot build housing for billionaires, then it obviously cannot build affordable housing for the working poor. Alex Schwartz, one of the leading housing scholars in the country, put it bluntly: "The lack of housing affordable to the lowest income renters reflects above all the inability of the private housing market to produce and maintain low-cost housing without public subsidy."[24]

But if the private sector cannot fix the housing crisis, who can? Most people probably assume the federal government builds public housing, and the solution to the crisis is simply for the government to build more. But the federal government hasn't built public housing for more than two decades. In its passionate embrace of neoliberalism, the Reagan administration cut the budget authority for the Department of Housing by 70 percent, and since 1996, through Republican and Democratic administrations, no money has been authorized for the construction of new public housing. In fact, quite the opposite: over the past twenty-five years, the United States has demolished scores of public housing projects, reducing the available supply of public housing by over a quarter-million units.[25]

If the free market cannot, and the federal government will not, solve the crisis of affordable housing, then what is the solution? In our neoliberal moment, policy makers in the United States have created an elaborate but fragile network of public-private partnerships in the hope that the private sector, if given the right financial rewards and incentives, will close the gap between demand and supply. A close study of these partnerships is a trip down a rabbit hole. The programs are enormously complex, with a bewildering array of initiatives, all governed by impenetrable statutory language and byzantine regulations. Though the details matter, it is vitally important not to lose sight of the elemental feature that sustains the entire apparatus. The United States will provide affordable homes only if—and as long as—someone in the private sector can profit by it. *If there is no private profit, then there is no low-cost housing.* With very rare exceptions, profit is the oxygen of U.S. housing policy.

The most widely known housing program is the voucher system. The government provides qualifying applicants with a voucher that they can use toward the cost of rent in any qualifying unit. This is not the same as public housing. It is government-supported housing in the private market; the renter pays a portion of the total rent and uses the voucher to make up the difference. This system—known as Section 8 for its place in the law that created it—is the largest housing subsidy program available to low-income Americans.[26] In researching this book, I grew accustomed to seeing small signs nailed to the walls of Olneyville townhouses and duplexes: "We accept Section 8." While the number of families in public housing has fallen since the early 1990s, the number receiving vouchers has grown significantly and now includes about 2.3 million households.[27]

It is not remotely enough. The need is far too great, especially in cities with the tightest real estate markets. Section 8 sets limits on the amount a landlord can charge for rent, and in tight markets, where landlords can get far more for their rentals than Section 8 will authorize, they have no financial incentive to participate in the program, especially since it also involves regulations and inspections to make sure the owner is keeping the unit up to code. So precisely where Section 8 housing is needed most, it is least likely to be available. As a result, eligible renters routinely wait years for a voucher, and most never receive one at all. In Los Angeles, the wait is more than a decade, and when the local housing authority announced in 2017 that it would accept new applications for the first time in thirteen years, it anticipated 600,000 applicants for 20,000 spots. But at least Los Angeles added families to its waiting list. In most places, the waiting list is so long and turnover so infrequent that cities have closed the lists indefinitely. In Providence, the list opened for six days in 2016, which was the only opening this century. The city has no current plans to reopen its list. In Olneyville and thousands of neighborhoods like it, Section 8 offers a vital lifeline for some low-income renters but is utterly irrelevant to most others.[28]

Section 8 has a second problem. Because comparatively few renters have Section 8 vouchers, their needs can typically be accommodated by the existing supply of homes. The voucher program therefore does not increase the total supply of below-market shelter. But what a distressed neighborhood like Olneyville truly needs is *more affordable housing*. For that, neighborhoods must rely on a different federal program, aimed at builders and developers. It is the Low-Income Housing Tax Credit, or LIHTC (pronounced "lie-tech"), a federal program created in 1986 and made permanent in 1993. LIHTC awards developers a stream of tax credits in exchange for constructing affordable housing. Each state is allotted a certain dollar value of tax credits, which the state then allocates to developers on the condition that they use the credits to build housing priced at below-market rates. The developers typically sell these credits to investors, who make a fixed return set by the government. Developers, including nonprofit community development corporations like ONE Neighborhood Builders, then use the money to fund the construction of housing. Provisions in the law allow the apartments to remain at below-market rates for up to thirty years.

Importantly, the investors lose the tax credits if the buildings are not built and maintained to a high standard. This provision helps guarantee that LIHTC housing will be not only more affordable but also safe and well maintained. For that reason, LIHTC housing has become an important part of community revitalization in distressed neighborhoods. Since the program's creation in 1986, LIHTC has helped finance the construction of over 2.5 million below-market units nationwide, making it by far the largest program in the country for the creation of affordable housing. Tax credit housing now accommodates about twice as many households as public housing.[29] So while Section 8 is a partnership between government and private-sector landlords, LIHTC is a partnership between government and a coalition of private and nonprofit developers, builders, and investors. But as with all things neoliberal, the system works only so long as someone in the private sector makes a profit.

Tax credit housing is what transformed Aleppo Street. After acquiring the vacant land and purchasing the derelict properties on and near Aleppo, ONE Neighborhood Builders used money that came principally from the sale of federal tax credits to build more than forty affordable apartments. Then, once the rentals had stabilized the area, the organization used a small federal grant to build an additional twenty single-family houses.[30] On space that was once an urban wasteland, ONE Neighborhood built more than sixty affordable apartments and houses that now provide safe, quality housing for several hundred Olneyville residents. When Frank Shea unveiled his plans for Aleppo Street, naysayers in Providence told him he would never get people to move to Olneyville. No doubt these were the same people who shook their heads in wonder at Jane Sherman's naïveté as she described her grandiose plans for Riverside Park. But as Shea proudly told me, a waiting list formed as soon as the homes became available. From the moment they were finished, they have been fully occupied, and Aleppo Street, with its sweeping views of Riverside Park, is now the most attractive street in the neighborhood.

What ONE Neighborhood did on Aleppo Street, it has since repeated at several sites throughout Olneyville. If you walk through
the neighborhood, you can always spot the organization's houses.
Whether stand-alone homes, side-by-side duplexes, apartments, or
townhouses, they are all built in the size and style of the neighborhood's historic mill houses. Newer by about a century, they seem to fit
in even as they stand out; they are invariably the nicest buildings on
any block. Since ONE Neighborhood's founding in 1988, it has built
nearly five hundred apartments and private houses, the majority of
which are in Olneyville, meaning that in a neighborhood of roughly
seven thousand people, more than a thousand live in safe, quality
homes, priced at below-market rates, built by ONE Neighborhood.[31]

I should mention one more aspect of single-family homes built
by ONE Neighborhood, if only because it radically departs from
the neoliberal model. Though renters greatly outnumber owners
in Olneyville, ownership lends stability to the neighborhood, since
owners have a greater stake in the area's long-term success. Homeownership also improves social conditions; neighborhoods that have
higher levels of homeownership tend to have lower levels of crime and
higher high-school graduation rates.[32] In addition, unless a neighborhood has a supply of homes for sale, it discourages those who can afford to buy a place from remaining. Ideally, therefore, a low-income
neighborhood will develop a mix of affordable rentals and private
houses, which is why Shea was eager to build a few houses alongside
the apartments on Aleppo Street.

But the houses constructed by ONE Neighborhood are not sold
under the same conditions as most homes. When most people purchase a house, they buy both the structure and the land beneath it,
and they can then sell them for whatever the market will bear. But
that means a house that was affordable to the working poor at one
point might no longer be affordable a few years later. That is what

happened to the modest three-flat that Frank's parents bought in 1962 in Jamaica Plain for $12,000, which their children sold for approximately $900,000. When this happens, the neighborhood is no longer affordable to the working class. The huge windfall for Frank and his siblings came at the expense of the neighborhood as a whole.

To prevent this pattern, ONE Neighborhood homes are always sold subject to a shared-equity agreement, which divides the rights and responsibilities of homeownership between ONE Neighborhood and the homeowner. When someone buys a ONE Neighborhood home, she purchases only the house and not the land beneath it, which is owned by the nonprofit. This allows her to buy a home at a substantially reduced price, but in exchange she gives up the right to sell the house for whatever she can get for it. Instead, if she chooses to sell, she will make a modest return (often agreed on at the time of her purchase) but can only sell at a price that is affordable to the next low-income purchaser. The new purchaser also buys only the structure, not the land.

This arrangement, which can last forever, helps a low-income neighborhood thrive by ensuring a stable supply of quality homes at below-market prices. This allows the neighborhood to resist displacement and gives an owner a greater commitment to the neighborhood by limiting her financial incentive to leave. In short, shared-equity ownership protects and enhances the neighborhood as a whole by abridging the rights of the individual homeowner.[33] These arrangements thus completely reject neoliberal orthodoxy. They deliberately decommodify land by taking it out of the private market and thereby place the interests of the group ahead of the rights of the individual.

Shared-equity arrangements of the sort employed by ONE Neighborhood, however, are not an inherent feature of LIHTC financing. On the contrary, most LIHTC homes are simply sold on the private market, albeit at a lower cost. LIHTC thus illustrates perfectly the neoliberal approach to neighborhood well-being. Rather than taking responsibility for building affordable housing, the federal government incentivizes private developers to do the work by guaranteeing a stream of income in the form of tax credits. As with all things neoliberal, the program works only because the national government has guaranteed that someone in the private sector will make money by providing an essential service to the poor. If the government does not provide the guarantee, then developers will not build the houses, and the poor will not have shelter.[34]

This points to the great limitation of LIHTC: it only works *to the extent* of the government's financial guarantee. To be eligible for a LIHTC apartment, a renter must earn no more than 60 percent of the area's median income. As calculated by the Department of Housing and Urban Development, the median income in 2019 for a family of four in Providence was $81,900, so that a family of four is not eligible for LIHTC housing if it earns more than $49,140. In practice, rents in LIHTC housing are set in the expectation that renters will earn approximately this amount. (The allowable limits are modestly lower for a smaller family, and modestly higher for a larger family.)[35] But in Olneyville, very few families of four earn $49,000 a year; in 2018, the latest year for which we have data, the median household income in the neighborhood was just over $27,000.[36] This means that the lowest-income residents who most need an affordable place to live cannot afford LIHTC housing.[37] Indeed, recent research shows that when LIHTC housing is built in distressed neighborhoods like Olneyville, it tends to raise nearby home values substantially and produce an influx of white, comparatively wealthier homeowners.[38] As the authors of one study put it, "LIHTC construction makes low income neighborhoods more desirable regardless of minority share."[39]

LIHTC funding works, therefore, but because the government does not guarantee a sufficiently large stream of income, the program does not work for a neighborhood's neediest residents, and it can contribute to gentrification by luring wealthier homeowners to a still-affordable neighborhood that appears to be on the rise.

This inequity can be fatal.

On January 2, 2018, eighteen years to the day after Frank Shea started at Olneyville Housing, Roland Colpitts called 311, the city service line in Providence. Colpitts complained, as he had several times before, that his apartment at 110 Bowdoin Street in Olneyville had no heat, no running water, frozen pipes, and exposed electrical wiring. The temperature in Providence that day sank to the low single digits and never climbed above 20°. Winds reached twenty miles per hour, and gusts were even stronger, making the temperature feel well below zero. The next day, the city sent a housing inspector to the home, who found that the owner-landlord had carved the building into a warren of private and semiprivate rooms, creating an illegal boardinghouse. The men and women crammed into the house had brought in space heaters to stave off the cold and propane torches to thaw the frozen pipes. One couple used a hot plate to stay warm. A tangle of extension cords ran from the second floor to an uncovered electrical panel in the basement. A single meter supplied electricity to the entire building, and power was intermittent. Garbage littered every floor, and "red hot" wires dangled over mounds of accumulated debris.[40]

"On the street, it's known as an emergency situation," Colpitts told the *Providence Journal*. "If you needed an apartment bad, really bad, you look on Craigslist and [this landlord] had an apartment ready. There was always people moving in and out." Colpitts and his girlfriend moved in because she needed an oxygen tank and could no longer live on the street. They gave the landlord a $200 security deposit and paid him $800 a month to live in a single room in a building with no heat and no running water. On the basis of what he saw, the housing inspector started condemnation proceedings, just as the city had done in 2015 when another set of tenants reported that the same landlord had failed to fix the same problems in the same building. At that time, the city had declared the house "unfit for human occupancy." The landlord had been cited for at least twenty-eight code violations since acquiring the building in 2010.[41]

Three days after the inspector toured the house, it burned to the ground. The fire began in one of the many illegal makeshift bedrooms in the early morning hours of January 6, 2018, and quickly consumed the entire wood-frame building. Colpitts woke up when he smelled smoke. When he opened a window, flames shot into his room. He slammed the window shut, woke up his girlfriend, and opened the bedroom door. "It was like someone hit me in the face with a shovel," he told a reporter the next day. "That's how fast the smoke came rushing in." By the time firefighters brought the four-alarm fire under control, it had destroyed not only 110 Bowdoin but the apartment buildings on either side as well. A dozen families lost everything they owned, seven tenants had to be sent to the hospital, and two firefighters suffered minor injuries when the roof of one of the buildings collapsed on top of them. One tenant, forty-nine-year-old Lucy Feliciano, died in the fire.[42]

Cecil Vega and her four children lived in a three-bedroom apartment on the third floor of 108 Bowdoin, next door to 110 Bowdoin. (Cecil spells her name without the final *e* but pronounces it *Cecile*.) In the middle of the night, her daughter came into her room and woke her. "Mommy, I smell smoke." Cecil opened her bedroom window, put her head out, and saw the fire next door. "Run!" she told her daughter. "Just run!" Pandemonium reigned as Cecil raced through the three-bedroom apartment, waking the children and hustling everyone downstairs as black smoke billowed up the stairwell. Cecil was the last to leave the building, carrying their dog in her arms as she stepped outside, barefoot, into the freezing January night. Almost immediately, the neighborhood rallied to the aid of those who had lost their homes. "Blankets, food, clothing. Everything," Cecil told me. The teachers at William D'Abate Elementary School organized a donation drive for Cecil and her children. So did ONE Neighborhood Builders. Even as the fire burned, a neighbor who lived across the street took Cecil's children into her home to escape the cold. It seemed like help came from everywhere. "People I didn't even know," she told me.

I met Cecil a little over five months after the fire. She is one of the lucky few in the neighborhood who has a Section 8 voucher. Unlike the landlord at 110 Bowdoin, who made no effort to maintain his building, Vega's landlord received enough in rent—from the tenants and the government—to keep the building in good shape. She had no complaints about their apartment. The utilities worked, the building was sound, the grounds were clean, and the place was safe. After the fire, she had to split up her family; she and her daughter stayed with cousins, and her two sons lived with their father nearby. After three months, a three-bedroom apartment became available in another Section 8 home in Olneyville, owned by the same landlord she had rented from before, and he offered it to Vega. "That was nice of your landlord, to offer you this place," I said when we met in her new living room. "As far as I'm concerned," she told me, "he's an angel sent from heaven."

Vega had been in the new apartment for just two months, and most of the walls were still bare. "We lost everything," she said. But she had recently hung her son's newly won certificates of academic achievement from middle school on the living room wall. After the fire, a crew from the city combed through the debris and found a single charred photograph, which they traced to Cecil and delivered to her at her job. The photograph shows her oldest daughter holding her baby cousin in her arms and kissing her cheek. Cecil keeps the picture in a manila folder. It is all that she and her family have left from their life before the fire. "But I didn't really care about that," she told me, "as long as my family was safe. The rest is just material stuff. The only thing I felt really bad about was, I lost my mother's ashes."[43]

Like the creation of Riverside Park, the struggle for affordable housing in Olneyville reveals both the promise and the limits of the current approach to neighborhood well-being, an approach I have called neoliberal precarity. On one hand, Aleppo Street, once desolate and dangerous, is now home to hundreds of Olneyville residents, all of whom live in safe, high-quality housing across the street from one of the most attractive parks in the city. The importance of this achievement should never be diminished. ONE Neighborhood Builders also took the necessary steps to guarantee that the housing will be priced below market rates for many years. And what ONE Neighborhood did on Aleppo Street, it has done throughout Olneyville, creating hundreds of units for residents who could otherwise never afford shelter of this quality.

On the other hand, the Bowdoin Street fire reveals the limits of this partnership. It cannot remotely satisfy the demand for affordable housing within a low-income neighborhood like Olneyville. It is not for want of tools that we still have illegal homes like the one that burned to the ground on Bowdoin Street, killing Lucy Feliciano. It is for want of will. We could build thousands of apartments like those on Aleppo Street, and they would repay the investment many times over. We could make these units affordable to the poorest renters and employ shared-equity arrangements that would protect against the market forces that drive poor people from their neighborhoods. We could solve the housing crisis. We simply refuse to do it.

Imagine a fire raging through a neighborhood. We know that

with enough water, we can put the fire out. No one would tolerate a government that said, "You can have this much water, but no more," knowing full well that it had allotted far too little to meet the need. Nor would we tolerate a government that said, "You can have all the water you want, but the fire company will not come to put it out unless it makes a large-enough profit." But that is what we have done with housing in the United States. The amount of financing available from the federal government is not nearly enough to bring the fire under control, and unless the government guarantees a sufficient rate of return, the private sector will not participate. These two conditions have always limited the capacity of neoliberalism to solve the housing crisis.

Cecil Vega credits her new apartment to "an angel sent from heaven," but Lucy Feliciano did not die for want of divine intervention. She died because she could not escape the smoke and fire that swept through the illegal boardinghouse where she had paid an exorbitant fee for the privilege of huddling in a freezing room without heat or running water. She died because the neoliberal approach to neighborhood well-being does not build remotely enough affordable housing, either in Olneyville or throughout the country. It *could*, if it were adequately funded, but it doesn't, because it isn't. The tools exist, but they lie unused for want of funding. Lucy Feliciano died because in the United States, housing is not a right.

The story of American housing policy is the story of a failed state, not as the term is used by political scientists, since that implies the state's inability to maintain authority over its people or project its authority to the world, but in the moral sense, in the sense of the state's ability to provide basic and essential services to its people. In that sense, which is the sense that matters, the United States is a failure. Routinely and unapologetically, it fails to meet the elemental needs of a large and growing fraction of its population. We have a housing crisis because the market cannot, and the government will not, produce enough affordable shelter. It is entirely within our power to fix the problem, but instead we rely on a neoliberal approach that we *know* is inadequate.

What Jane Sherman did for Riverside Park, ONE Neighborhood Builders did for housing. And just as the creation of the park revealed one set of problems with the world of neoliberal precarity, so the transformation of Aleppo Street revealed another. The park depended on elites from outside the neighborhood who tapped their personal and professional networks to extract money and political commitment, producing a collaboration that cannot be replicated and that leaves the neighborhood with no enduring skills or expertise. Aleppo Street relied on a deliberately inadequate stream of federal funding that predictably fails the neediest residents while it increases gentrification. In both cases, the current approach worked about as well as it could and made the neighborhood substantially better for at least some low-income residents. Yet in both cases, it also made Olneyville much more attractive to wealthier newcomers. And in both cases, the lesson is the same: the current approach can do much to transform a distressed neighborhood but cannot protect it.

5

You Can't Build a House with Handcuffs

You're well on your way to gentrifying a distressed neighborhood and displacing the people who call it home. Residents and their allies cleaned up that brownfield and rebuilt some of the housing. Their labors, which didn't cost you a cent, made the neighborhood much more attractive to wealthier newcomers, who have already moved into the LIHTC housing. But you still have much more to do. What about the police? Policing that is ineffective, brutal, or corrupt isn't good for anyone.

Frank Shea left ONE Neighborhood Builders in 2015 for a similar position in Boston. Looking back on his decade and a half in Olneyville, he could point with pride to the striking transformation of Aleppo Street. In the long run, however, that was probably not his most important contribution to Olneyville's well-being. Before he left, Shea revitalized the Olneyville Collaborative, a collection of nonprofit and community organizations that work in the neighborhood. The collaborative had become inactive before Shea's arrival in 2000, but he knew the neighborhood could not succeed without it. A place like Olneyville, where the needs are great and the funding is meager, can ill afford the competition that so often bedevils the nonprofit sector. No one wins, least of all the residents, when groups that ought to be allies begin to brawl over limited resources. Shea reinvigorated the collaborative to prevent this kind of conflict and to foster a spirit of cooperation and mutual respect among the many groups in Olneyville.[1]

One of the first groups he invited to the table was the Providence Police Department.

When I met him, Dean Isabella was a captain in the Providence Police Department. He was born and raised in Olneyville during the 1970s and '80s, when crime and drug use were at their worst. "I grew up in that very area, around Aleppo Street. I can still remember it like it was yesterday, coming down the stairs of my apartment building and coming up to a guy shooting up, lying at the foot of the stairs. I stopped, and he looked at me and I looked at him, and I'll never forget, he said, 'Whatever you do, kid, don't step on my needle.' So I jumped over him and ran outside to play. That was just life. That was where I lived."[2]

Isabella and his five brothers and sisters were raised by his mother. As the working poor have done for decades, they moved frequently, relocating every time the landlord raised the rent, always on the hunt for a cheaper place but always staying in or near Olneyville, where rents were low. Watching his mother struggle left a mark on Isabella. "I'll tell you one thing, you always hear people say poor people are lazy. Don't tell me poor people are lazy. My mother worked three jobs — single mom, six kids. When she came home, she couldn't take her shoes off because her feet were swollen, and if she took her shoes off, she'd never get them back on to go back out for her third job at an overnight diner. I never knew *anyone* in my neighborhood that was lazy."

In Isabella's Olneyville, children did not grow up with dreams of rising into the middle class. "Growing up," he told me in our first meeting, the thought of going to college "was a *pipe dream*. Owning your own house was a *pipe dream. Nobody* did that." Instead he followed one of the few paths that has always been open to working-class kids, at least if they are white.[3] He joined the Air National Guard at eighteen, not because he particularly wanted to serve in the military but because he thought it would help him land a position in law enforcement. He was honorably discharged at twenty and immediately started applying to police departments.

"I always knew I wanted to be a cop," he told me. "I don't know why. It wasn't like a family member or friend was a cop, or anything like that. But it's something I always wanted." He can still remember the day he learned he would become a Providence police officer. "I was twenty-two, and I remember getting the letter, and I knew what it was. I was still living at home, and I went into my bedroom and sat on my bed, just holding the letter. And I must've waited three minutes before I opened it. And then I saw 'Congratulations! You've been accepted into the Providence Police Department Academy.' And that was my life. And here I am thirty years later." He took a handkerchief from his pocket and dabbed his eyes. "You can see I'm getting emotional about it."

We were sitting in his small, square corner office on the ground floor of the Providence Public Safety building, talking about changes since he joined the force. "When I started in 1987," he said, "we were a traditional force—." "What do you mean by that, traditional?" I interrupted. "Quasi-military," he answered. At the most innocuous level, to describe a police department as quasi-military, or more commonly paramilitary, is simply a comment on its organizational structure. A paramilitary organization has a rigid, top-down command style that employs a centralized decision-making process and grants relatively little discretion to officers in the field. Used this way, the term is not limited to policing. I have met with prison officials, for example, who have described their organizations as paramilitary.

Still, in my experience, few officers use the term simply to describe their department's organizational style. A police department that adopts a military-style command structure generally also embraces a military mind-set, which shapes the department's orientation to the outside world. Like most cops, Isabella meant "quasi-military" in this broader sense. He told me that when he started in the Providence Police Department, there was "a militaristic attitude reflected in everything you did." This attitude comes naturally to many police officers, likely because roughly one in five cops, like Isabella, have previously served in the military.[4]

This is where things can go awry. Being a soldier, like being a police officer, can undoubtedly be an honorable profession. There is, or can be, great virtue in the selflessness and commitment to public service at the heart of both military service and law enforcement. Yet obedience to all lawful commands is an essential aspect of military discipline. Soldiers do not expect to have their orders questioned and generally do not pause to explain or justify those orders to the soldiers and civilians under their command. But whatever wisdom lies in demanding unquestioning obedience in the military, it takes no particular imagination to see the effect this mind-set can produce in a domestic police department.

Many departments actually *encourage* officers to see themselves this way, and to model their behavior not just on the common soldier but on the soldier who is constantly threatened by enemies. This warrior model of policing originated out of a legitimate concern for officer safety, but that concern soon mutated to dwarf all other considerations, including neighborhood well-being.[5] Under the warrior mind-set, officers are conditioned to believe that threats swirl around them unseen—behind them while they wait in line for coffee, in their blind spot as they hand a summons to a pedestrian, or in the seemingly innocuous encounter with a resident complaining about garbage in the street. The warrior cop is trained to view every civilian interaction "as a potential deadly-force encounter."[6] Some veteran police trainers are perfectly explicit on this score. "Develop that warrior mindset that allows you to become the predator and not the prey," one senior trainer advises. "Remain humble and compassionate; be professional and courteous—and have a plan to kill everyone you meet."[7] That, Isabella told me, is how Providence trained its officers to see the world.

At the risk of stating the obvious, an officer who demands unquestioning submission—who views every encounter as potentially life-threatening and has a plan to kill everyone he meets—may not respond well to disobedience. A two-year study of police brutality in New York City identified a single "iron and inflexible rule": to defy the police is to risk retaliation, "commencing with a summons, on up to the use of firearms."[8] Inevitably, this will sometimes lead to tragedy. In July 2016, a Minnesota police officer shot and killed Philando Castile during a traffic stop. It turned out the officer had undergone a popular training program called "The Bulletproof Warrior," put on by Calibre Press, a publisher of police-related books and videos. The course included videos of officers being shot, and booklets recommending that police use more force than necessary, even preemptively. One booklet read, "Myth: The officer must use the minimal amount of force necessary to [effect] their lawful law enforcement objectives."[9] Dave Grossman, a former professor at West Point who has taught the course to thousands of officers, emphasizes to attendees that their job is to fight violence with greater violence; once an officer is properly prepared, "killing is just not that big a deal."[10] (The officer who shot and killed Castile was acquitted at trial.)

The Calibre Press course is not an aberration. In 2015, the Police Executive Research Forum found that police departments spend an average of 8 hours training officers in conflict de-escalation and 129 hours training them on weapons and fighting. With training so disproportionately focused on the use of force, it is easy to imagine how routine police encounters can escalate into brutality.[11] In July 2015, a Texas state trooper named Brian Encinia stopped twenty-eight-year-old Sandra Bland for failing to use her turn signal. He approached her car on the passenger side and, after a brief exchange, ordered her to put out her cigarette. Bland, who certainly had the right to smoke in her own car, refused. Now she had done something far worse than fail to use her turn signal. She had questioned Encinia's authority, thus committing what the police scholars James Fyfe and Jerome Skolnick call "a police cultural crime."[12]

Encinia ordered Bland out of her car. She refused this order as well. He then returned briefly to his squad car, walked to the driver side of Bland's car, opened the driver's door (which Bland had left unlocked), and tried to drag her from her car. When this failed, he pulled out his stun gun and shouted at Bland, "I'm going to light you up!" Bland got out of her car. Encinia restrained her hands behind her back, placed her in the back of his squad car, and took her to jail, where she was unable to post $5,000 bail. Days later, she was found hanging in her jail cell, dead.[13]

In one sense, the death of Sandra Bland is atypical. Police interact with millions of people every day, and the overwhelming majority of those encounters do not end in violence or tragedy. Yet in another sense, the interaction between Bland and Encinia is merely one extreme of a typical police-civilian interaction. Many officers trained as warriors deliberately adopt an attitude that demands obedience. The police call it "command presence." Officers communicate—through the words they use, the tone they adopt, and their nonverbal cues—that refusal to obey is not an option, and if a situation escalates to violence, the officer will never be on the receiving end. As one veteran trainer put it, "Command presence communicates to everyone present that you are in charge—not just now, but right the fuck now. Not just sort of in charge, but totally and completely fucking in charge."[14]

Command presence may safely calm some volatile situations, but when police are trained to view every situation as potentially life-threatening, they become quick to adopt an authoritarian persona. In practice, most people will dutifully obey an armed officer who demands that they act in a certain way, particularly when stories of lethal police violence are so well known. But no one takes well to being ordered around like a circus animal, and while Officer Right-the-Fuck-Now may secure compliance and drive away from a scene convinced he handled it correctly, he leaves behind a slow boil of resentment and barely contained hostility.

In this environment, brutality is inevitable. As Skolnick and Fyfe observed, "The *war model* of policing encourages police violence." No less than soldiers, cops at war need an enemy. The enemies they select "are found in inner cities and among our minority populations. There, in a country as foreign to most officers as Vietnam was to GIs, cops have trouble distinguishing the good guys from the bad." Before long, "*everybody* becomes suspect in their eyes. The community and the police become alienated and distrustful of each other, and incidents like the [Rodney] King beating occur more frequently than we would like to think" (all emphases in original).[15]

So it was in Providence. In 1991, the Department of Justice released a study of police brutality nationwide. The DOJ examined more than 15,000 complaints it had received between 1985 and 1990 and found that just 187 departments, out of more than 17,000 police departments nationwide, accounted for almost half of all complaints. When other researchers examined the data, they found that complaints were not distributed evenly across these 187 departments. Some were practically off the scale. The entire study found only 0.5

complaints per 1,000 officers in New York City, only 1.2 per 1,000 in Philadelphia, and 2.1 per 1,00c in Boston. In New Orleans, meanwhile, there were 26 complaints per 1,000 officers, a rate fifty-two times higher than in New York. The list of 187 jurisdictions contained only 4 from New England. Three were in Rhode Island: Providence, East Providence, and Pawtucket. The Providence Police Department generated more than 21 complaints of police brutality per 1,000 officers, ten times higher than the rate in Boston and forty-two times the rate in New York. Providence had a higher rate of brutality complaints than any major city in the country apart from New Orleans and El Paso.[16]

Violence is not the only product of the warrior mind-set. Experienced cops will tell you they cannot do their jobs without the support of the residents where they patrol, who know far better than the police who is doing what, who makes the neighborhood better, and who makes it worse. But what resident will support a force that views a neighborhood's members not as partners but as threats? The warrior model of policing pounds a sharp wedge between the police and the people they serve, driving them further apart with each belligerent glance and resentful sneer, and destroying the trust on which relationships are built. This is certainly what happened in Providence, where the department grew estranged from the very community it was trying to serve. None of the animosity or estrangement comes as a surprise to Dean Isabella. "If you train to become an occupying army," he said, "guess what you'll become?"

Yet Providence also had a problem that other cities did not. Though many departments encourage their cops to see themselves as warriors, only Providence had Buddy Cianci as mayor. For years, Cianci had used the department, as Isabella put it, "as his own private gestapo." Many of the city's most senior police officers secured their positions not on merit but because they contributed generously to Cianci's slush fund, a fact that was widely understood among the rank and file. A subsequent investigation found that corruption in the department was "rampant," and the promotion system was marred by years of "cheating, collusion, and political influence."[17] Nor was the corruption confined to promotions. An FBI investigation uncovered "significant problems . . . relating to testing, recruitment of candidates, the handling of seized vehicles, possible insurance fraud and potential bribery."[18]

And this corruption does not even count the senior officers who turned a blind eye as they escorted Cianci on his misdeeds, including being present while he attacked a man he thought was having an affair with his estranged wife (the assault that led to the first of Cianci's two felony convictions). One of the witnesses against Cianci at his second trial, this time for corruption, was Urbano Prignano Jr., former chief of the Providence Police Department, who testified under a grant of immunity about the culture of illegality and impunity in the department. Prignano resigned in 2001 as the federal probe into police corruption exposed payoffs for promotions and case fixing. This was not a command staff especially concerned with legal niceties.[19]

The transformation of the Providence Police Department began on Monday, June 24, 2002, the day a federal jury convicted Cianci of racketeering conspiracy. In September, he was sentenced to serve five years and four months in prison. State law required that he resign from office upon being sentenced, and that November, Providence elected David Cicilline as its new mayor.[20] When he took office, Cicilline discovered a police department that was thoroughly demoralized and at war with the community that most needed its help. The department "really was the king's army," he told me one morning in his Pawtucket office. By the time we spoke, Cicilline had been elected to represent Rhode Island in the U.S. House of Representatives, but he still vividly recalled the chaos in the Providence Police Department when he became mayor. "The whole system was so corrupt, and everyone knew it. I knew we needed a major restructuring." Cicilline reached out to Meg Curran, then the U.S. attorney in Rhode Island, who put him in touch with Andy Rosenzweig, a legendary investigator, first with the New York Police Department and later with the Manhattan District Attorney's Office. Rosenzweig gave Cicilline a name: Dean Esserman.

At first glance, nothing about Dean Esserman says cop. Rumpled, hulking, and jowly, he looks like a traveling salesman who just spent eight hours behind the wheel of a car too small for his frame. But all of that is forgotten after the first few sentences. We were talking over dinner one evening about the seemingly intractable challenge of restoring trust between neighborhood residents and a police force trained to be an occupying army. Scores of writers, from historians and journalists to sociologists and political scientists, have pondered this challenge, analyzing and scrutinizing every twist in a tortured road. But Esserman is a practical man, more interested in results than speculation, with an uncanny knack for distilling complex ideas into simple bromides. "It's hard to hate up close," he told me. And with that, like a walking Zip drive, he had compressed an entire library into a single sentence.

Esserman followed an unconventional path to policing. His father was a physician and a professor of medicine at New York University, his mother a child psychologist who started and later ran a market research company focused on children. Born and raised in Manhattan, politically liberal and culturally a New Yorker, Esserman attended Dartmouth College and NYU Law. After graduating in 1983, he landed a coveted job as an assistant district attorney in Brooklyn and later became a federal prosecutor, handling narcotics and organized-crime cases. Thousands of young lawyers have climbed the rungs from Ivy League college to top-tier law school to a stint with the local DA before moving on to become federal prosecutors, finally coming to rest as judges or partners in prestigious law firms. They are the 1 percent of the American Bar. But you can probably count on one hand the men and women who follow this path and end up a cop.

For Esserman, the turn came when he left the DA's office in 1987—the same year Dean Isabella joined the Providence Police Department—to become general counsel for the New York City Transit Police, where he had interned during college. There he came to know and work with Lee Brown and Bill Bratton. Brown, an early champion of what he called neighborhood policing—now known as community policing—became police commissioner (head of the department) in 1990, the same year Bratton became chief of the Transit Police.

In 1991, Bratton and Brown helped Esserman land a position as deputy chief of police in New Haven, Connecticut, making him a senior officer on a force of several hundred even though he had never walked a beat, never made an arrest, and never been certified as a peace officer. Policing can be an insular culture, with a strong sense of camaraderie that develops from the shared experience of doing each day what many people cannot understand and do not appreciate. "The most important academy in any department is the police locker room," Esserman told me once. That is why he attended the New Haven Police Academy in 1992, not as a new recruit but as a thirty-nine-year old senior commander, to prove to himself and those under his charge that he could be a cop, not just some liberal New York lawyer. He took over his own department in 1993, becoming chief of the Metro-North Rail Police Department, and in 1998 was named chief of police in Stamford, Connecticut.

That's where he was in 2003 when newly elected Mayor Cicilline asked to meet with Esserman about the position in Providence. At the time, the city was suffering. Murder and other violent crime rates remained stubbornly high, despite having fallen for at least a decade throughout much of the country. The corruption, brutality, and cronyism of the Cianci era had taken a toll on the department, and the relationship between the Providence police and the neighborhoods where protection and stability were needed most had reached a nadir. Distressed neighborhoods like Olneyville were trapped: they wanted security, but not the sort of policing that Providence had always provided, which failed to solve the problems that mattered and inflamed tensions that needed to be soothed. Originally Cicilline just wanted Esserman's advice: Who could get results without alienating neighborhood residents? In the end, Cicilline decided the answer was Esserman himself.

Cianci had corrupted nearly the entire senior command staff by opening the top positions to those who contributed most lavishly to his slush fund. Esserman replaced almost all of them, firing a few and shuttling the rest into miserable assignments from which they eventually resigned. Then he set about promoting an entire cohort of officers who embraced a radically different vision of policing. Like any leader, Esserman wanted results, but he defined success differently from his predecessors. In a traditional department, police tend to measure success by counting the things that are easiest to tally: How many 911 calls did they respond to last month? How many arrests did they make over Labor Day weekend? Did the number of shootings go up or down last year? But every experienced officer knows that these stats can be misleading. Suppose a department declares war on corner drug dealers. Six months later, it dutifully observes a dramatic decline in 911 calls complaining about corner drug dealers. The department's public relations machinery will surely turn these results into a nice story on the ten o'clock news. But if calls have declined because neighbors have been brutalized by the police and fear that law enforcement will make things worse rather than better, then the statistic means nothing. Or if success results from a temporary surge in policing or the arrival of especially cold weather—conditions that everyone knows won't last—then three weeks after the extra patrols end or the warm weather returns, the dealers will be back, and the neighborhood residents will lose even more faith in the police, whom they will see as more interested in misleading numbers and phony publicity than neighborhood well-being.

But Esserman did not define success by the quick fix and the easy victory. Influenced by Lee Brown and his idea of neighborhood policing, Esserman viewed success as a healthy, safe, vibrant neighborhood. In other words, he defined success precisely as residents do. To Esserman, and to the most progressive, reform-minded chiefs and police officers around the country, the question—first, last, and always—is simply whether policing helps the neighborhood get better. Policing that helps residents achieve that goal is good, and policing that doesn't, isn't. A cop who helps fix a problem without making an arrest but with the enthusiastic support of neighborhood residents is better than a cop who fixes a problem simply by taking out her handcuffs, and vastly better than a cop who is naive enough to think an arrest alone will solve a structural problem. Sometimes an officer needs to make an arrest, but the arrest is only useful if it makes the neighborhood a better place. "I tell my people it's not a fair world," Esserman used to say. "You produce bad results, you got problems with me. You produce good results but alienate the community, you got a problem with the community. You've got to answer to both."[21]

Some might find this remark perplexing. How could an officer get "good results" but still alienate the community? The best illustration I saw of Esserman's approach took place not in Providence but eight hundred miles away, in Cincinnati.

Maris Herold was a captain in the Cincinnati Police Department. We were sitting at a conference table on the second floor of CPD headquarters, talking about policing. She was describing an operation she ran when she was commander of the Fourth District, which includes some of the most troubled parts of the city. Police had wired confidential informants and sent them into Sulli's, a small market at the corner of Washington and Sullivan Streets, where they had no trouble buying crack and heroin from employees. One time an informant bought a gun. It was all on tape. (I have changed the name of the market and the streets.)

"This was a really bad store," Maris told me. "We'd been watching it for months. I could've busted them easy. Shut them down. Taken their liquor license." The police run thousands of operations like this all over the country, and if Maris had followed this well-traveled path, the outcome was all but certain. The market would have been shut, leaving an empty shell in the heart of the neighborhood. The men who sold the drugs would have been arrested, charged, and convicted of felony distribution of narcotics. They would probably have pleaded guilty in exchange for reduced sentences and ended up in one of Ohio's twenty-eight prisons. Because shootings and murders in Cincinnati were up from the year before, the arrests and the prosecution would have made good headlines.

Sulli's sits in an old part of the city, a couple miles north of downtown. Maybe ten parking spaces line the outside of the store, like teeth on a comb. Drivers can pull in easily, but most people walk to the store from the surrounding neighborhood. Two-thirds of the neighborhood residents are Black. The blocks nearby have a mix of small single-family houses built in the last few decades and larger ones with broad porches that date to the interwar years, along with a few red-brick Victorian townhouses that are even older. For the most part, the houses are well kept; only a few are noticeably run-down. In my visits, I saw no graffiti to speak of, and no trash in the streets or yards. The area is poor, but no one could fairly use a word like *blight* to describe it.

The store is tiny, about a quarter the size of the typical gas station and market you see at a highway rest stop or suburban strip mall. It has two narrow aisles crowded so close together that when a young man walked by me, we both had to turn sideways so that he could pass. You can walk the length of each aisle in six or seven steps. The shelves and coolers are crammed from floor to eye level with chips, cookies, soda, cereal, and fruit drinks, high in sodium and processed sugar but not very nutritious. Beer and malt liquor are sold in six-packs, singles, and quarts. Cigarettes and rolling papers are behind the counter at the front, just inside the door. One of the owners stands in a small cubicle behind bulletproof glass while the other deals with the customers at the register. At first glance, it doesn't look like the place adds much to the neighborhood.

But take a closer look. Alongside the colas, chips, and crackers, within reach of the beer and single-wrapped pieces of beef jerky, are milk, orange juice, oatmeal, and fresh bananas; diapers, paper towels, soap, and light bulbs; extension cords, toilet paper, and pencils; behind the counter, next to the cigarettes and cigarillos, are pain relievers, cough syrup, and eye drops. The man who passed me in the aisle bought a twelve-ounce carton of orange juice and a quart of Magnum malt liquor. Sulli's is the only store of its kind for blocks around. "This is where poor people shop in the city," Iris Roley told me. She is a community activist and part of a small but tenacious group of activists and religious leaders in the city, mostly Black, who have spent much of their adult lives trying to reform police practices in Cincinnati. "For a lot of people, they can't just hop in the car and drive to Trader Joe's. I told Maris, 'You can't close that store.'"

And she didn't. Rather than shut it down and arrest a few more low-level dealers, which would have accomplished little, Maris decided to make it a better and safer market. The employees who sold the gun and drugs were fired by the owners, who have a different commitment to the store and a greater incentive to keep it on the right side of the law. A prominent sign posted on an exterior wall warns, "No Drug Dealing. No Loitering. No Prostitution. We Call Police." A large streetlight, newly installed, now shines down on the corner from its position at the edge of the lot. Nobody was arrested, and nobody was prosecuted. No property was forfeited, no licenses lost. In short, the Cincinnati Police Department, in conjunction with the city, fixed the market by changing it. They improved the guardianship and altered the physical conditions to provide a more secure environment and disrupt the opportunity for crime and disorder. In doing so, they honored the wishes of residents and preserved an important neighborhood resource. "Now it's a good store," Maris said. Iris agrees. "It's an important part of the community." Maris calls the operation at Sulli's a complete success. "What we did at Washington and Sullivan, I want to do throughout the whole city of Cincinnati." Dean Esserman would concur.[22]

To achieve his focus on neighborhoods, Esserman reorganized the entire Providence Police Department. Before his arrival, the department had operated much as any urban force does. Policing was almost entirely reactive: crime happens, victim calls 911, cop appears. Officers on the beat responded to calls for service, wrote reports, and passed the reports on to their superiors, who deployed what few investigative tools they had. Sometimes they closed a case.[23] Providence cops made little effort to prevent crime and paid even less attention to the environmental or structural conditions that gave rise to a particular problem or made crime more prevalent in one location than another. Cops on the beat did not think it was their job to prevent crime or understand these conditions. That doesn't mean they were blind to neighborhood realities; any officer worth her salt will recognize that a distressed neighborhood needs a lot more of almost everything—housing, parks, streetlights, bus stops, free clinics, grocery stores—but until Esserman arrived, these conditions were not a cop's concern. Until a crime was reported, police did not consider it their role to intervene. Cops in Providence did not think they were responsible for making a neighborhood better.

This concept of responsibility is what Esserman changed. He and his new command staff divided the city into nine districts and, in each one, named a lieutenant as district commander. This organizational change was implemented only to facilitate the more important philosophical change: each district commander became responsible for making the neighborhoods within his district better places to live, a responsibility he would then communicate to the officers under his command. Decision-making responsibility no longer flowed downward from police headquarters to the districts. It originated within the districts and was shared by the district commander and the officers on the street—precisely the opposite of a paramilitary structure. Rather than simply reacting to crimes that had already occurred, officers were charged with identifying site-specific crime and disorder problems and designing creative, proactive solutions that would make the neighborhood a better place to live and work.

As simple and obvious as this approach may sound, it was a radical reform. Once a commander becomes responsible for solving crime, she quickly realizes how little she can do—how little any cop can do—to solve it. As a cop, she often simply lacks the tools necessary to create enduring solutions. She can make arrests all day long, but you cannot arrest your way out of structural or environmental conditions. You cannot build a house with handcuffs. Yet Esserman had made it the commander's job to solve these problems, and if she failed, she would have to answer to him. An economist might say that Esserman shifted the cost of failure from the residents alone to the residents and the district commanders. By necessity, this aligned the commanders' interests with those of the residents and forced commanders to open themselves up to partnerships—to become less insular and more receptive to collaboration with people who did not wear a badge or carry a gun.

To succeed under Esserman, commanders needed to become intimately familiar not simply with the crime and disorder problems in their districts but with all the resources that might be brought to bear to solve those problems. These resources could come from anywhere, including residents, businesses, churches, community organizations, and municipal departments. Most of all, however, the commanders needed allies. To forge alliances, they had to get out of their silos and become part of the neighborhood, a proximity that not only gave them insight into the problems they were charged with solving but humanized them to the neighborhoods they served—because, as Esserman understood, it's hard to hate up close. Under Esserman, district commanders became place-based problem solvers and the most important figures, after Esserman himself, in the redesigned Providence Police Department.

Like any strong leader, Esserman has plenty of detractors, and I spoke with many of them. To put it gently, he was not an easy boss. A story still makes the rounds about a talk he gave to the officers under his command when he became the chief in Stamford. "I want you up all night thinking about this joint," he said. "I want you back in the morning before you need to be because you can't wait to get to work. I want you thinking about your job. I want you dreaming about it. I want it to hurt your marriage. I want it to be your jealous mistress. I want it to be your obsession."[24] Stories still swirl of Esserman publicly belittling senior officers, humiliating them in front of their peers or berating them for not performing up to his impossibly high standards. Even his supporters admit he could be cruel to those who did not measure up.

Bob McKenna thinks that maybe Providence needed someone like Esserman. A veteran cop who moved into academia and worked closely with Esserman to reshape the Providence Police Department, McKenna told me he thought the department would have chewed up and spit out a softer boss. It took someone like Esserman to transform a department so hidebound by the paramilitary tradition and warped by corruption. David Cicilline agrees. "We needed a bold disruptor. He was confronting generations of corruption that was very resistant to change." Whether someone else could have done the job without his heavy-handed abrasiveness is something we will never know.

Esserman left Providence in June 2011 after reporters discovered underage drinking at his home. Apparently guests had snuck in alcohol at his teenage daughter's graduation party, and Esserman took the blame.[25] Shortly after, he was hired as the chief in New Haven, Connecticut, but resigned under pressure in 2016 after losing his temper in public once too often.[26] Now out of policing, he lives in Providence and works as a senior consultant for the Police Foundation, a leading police research and training organization. We met often to discuss his long career, his philosophy of policing, and the transformation of the Providence department. Over dinner one night, I asked him about his well-deserved reputation as an unforgiving taskmaster. His response surprised me.

"I regret all that," he said. "It was wrong. I was wrong. I think I wanted to show that I was a real cop, that I could be tough." Because he had such an untraditional background for a police officer, he thought he needed to show he could be even more hard-nosed than the longtime cops he needed to impress. Bosses like Esserman inspire extreme opinions, and people either loved him or hated him. Those who hated him had good reason, but one thing is certain: he eliminated a deeply embedded culture of corruption and lawlessness by kicking paramilitary cops out of their silo and forcing them to work with and for the residents they served.

Some scholars describe the transformation made by the Providence Police Department as a shift from paramilitary to community policing. I have deliberately avoided the label *community policing* because it is so overused as to have lost almost all meaning. At the most general level, community policing is simply a directive for a department to take the neighborhood's views into account when formulating police strategies. It is a philosophy rather than a technique and does not compel the police to adopt or abandon any particular strategy or organizational style. Most police departments now claim they practice community policing but have changed nothing about how they police. Many have not even abandoned the paramilitary structure or the warrior mind-set.[27]

In many departments, a handful of officers are assigned to a "community policing unit." This may be the worst of all worlds. Isolating the unit from the rest of the force not only limits the unit's effectiveness by starving it of resources but communicates to the rest of the department that they are *not* engaged in community policing.[28] Before Esserman's arrival, even the Providence department had a community policing unit, under the direction of Paul Fitzgerald. Fitzgerald told me the leadership left him and his team alone to do what they wanted, and although they accomplished some good things, it was a far cry from the root-and-branch approach Esserman adopted. Fitzgerald also told me he was not surprised at all by the 1991 DOJ study that found the Providence Police Department had one of the highest rates of police brutality in the country. Creating a community policing unit is no panacea.

If we insist on giving Esserman's approach a name, I would describe it as *neighborhood-aligned policing*. His philosophy represents a deliberate determination to align the success of the police department with the success of the neighborhoods where police are most needed. He forced the Providence Police Department to abandon the insular warrior mentality and adopt a collaborative, transparent approach built around long-term partnerships and relationships between the police and the people they serve, who collectively join hands to solve crime and disorder problems so as to create a thriving neighborhood.

The first test of Esserman's approach took place in Olneyville.

6

"What Was Done, Was Done Right"

To say, as Dean Esserman did, that his officers had to answer both to him and to "the community" sounds good in principle, but it raises as many questions as it answers. Which community? Who speaks for the neighborhood? Residents do not always speak with a single voice, and nonresidents may not have the same idea as residents of what the neighborhood needs. In a neoliberal moment, when local government cannot survive without luring outside investment, do the police listen to private developers or low-income residents? When gentrification begins, do the police align themselves with white newcomers or with the people of color who have lived in the neighborhood for years? There is no aspect of neighborhood well-being where disparities in power can have a more profound impact than in policing.

For many years, Providence residents who paid only casual attention would have described Olneyville as a "bad" neighborhood. When Jane Sherman told friends she was developing a park on the old Riverside Mills site, they thought she had lost her mind. *"Olneyville?"* they said. "They'll steal everything." But people willing to look beneath the surface have long known that even in the most distressed neighborhoods, the vast majority of crime is committed by a very small number of offenders. Violent crime is particularly concentrated. In Boston, for instance, a study showed that three-quarters of the gun assaults and one-half of the homicides between 1980 and 2008 were committed by fewer than 1 percent of the city's youth (aged 15–24), and most of these young men (and some women) had long criminal histories.[1]

Likewise, every cop who has ever walked a beat can tell you that some places are the site of far more crime than others. Recent research has shown that the connection between crime and place is even stronger than we suspected. In the same Boston study, researchers found that nearly nine in ten street segments and intersections over the entire twenty-eight-year period experienced zero firearm-related assaults that caused an injury. A little more than half of the remaining spots had only one such incident. Most of the gun violence in Boston during this period occurred at fewer than 5 percent of the street segments and intersections.[2] Additional research has extended these findings to other cities and other types of crime. Year after year, roughly half the crime in a city takes place at only 4 percent of the street segments, a quarter occurs at about 1.5 percent of the segments, and the great majority of segments are entirely crime free. (A street segment is the area on either side of a street between two street corners, but sometimes the troublesome spot is just a single address, a dot on the map.)[3]

So it was in Olneyville when Riverside Mills was a junkyard and Aleppo Street was a wasteland. Though the neighborhood had one of the highest crime rates in the city, most of its residents spent all their time on the right side of the law, and most places were perfectly safe. The lion's share of the trouble was concentrated within a few blocks around Aleppo Street, an area that represented only 3 percent of the entire neighborhood but accounted for nearly one in six calls to the police. Yet even this picture vastly overstated the dispersion of crime in the neighborhood; within this tiny box, crime was concentrated around four spots: two houses, one business, and the abandoned area along Aleppo Street.[4]

When crime and disorder are so tightly concentrated, it is tempting to think police can simply saturate those areas with patrols and arrest their way to a solution. But cops and criminal justice scholars have long recognized that arrests alone rarely work. They may provide a temporary fix by clearing high-crime places of criminal actors, but the police do not have the resources to maintain that saturation forever. If we cannot change the underlying conditions that attract crime and disorder, the same locations tend to stay "hot" year after year, regardless of the number of arrests. Of course, different places present different crime-control challenges. Sometimes the problems at a spot are physical, like dark alleys or obstructed views; sometimes they are personal, like the poorly managed corner market in Cincinnati. But unless you solve the problems, you will never fix the place.[5]

Occasionally, a place-based problem is simple to solve. I was once visiting with neighborhood activists in Ranier Beach, a low-income neighborhood on the south side of Seattle, who told me that groups of high school students got into fights every day after school, always at the same place. Sometimes the fights were quite violent. The activists were determined to find a solution that did not involve saddling more young people with criminal records, and when they took a closer look, they learned that the fights typically involved kids from two different neighborhood high schools, both of which released their students at the same time every afternoon. Students walking home converged on the place at the same time, and taunting sometimes escalated to violence. The activists persuaded one school to start and end its day a half hour earlier, so that its students had cleared the area by the time those from the other school arrived. Problem solved.[6]

In other places, however, the problems require more elaborate interventions. That is why the efforts of Jane Sherman and ONE Neighborhood Builders at Riverside Park and Aleppo Street were so important: they had the potential to change the conditions that made the area criminogenic. How could the Providence Police Department help? The solution was designed by Hugh Clements. Born and raised in Silver Lake, a neighborhood next to Olneyville, Clements is a long-time veteran of the department. He was an early champion of Dean Esserman's shift away from the paramilitary style, and in 2003, Esserman named Clements the commander of District 5, which includes Olneyville. It was Clements who conceived the police strategy at the site. In an interview in 2008, after he had risen to become chief of the department, Clements said that the most important challenges, particularly in the beginning, were "the blatantly obvious quality-of-life issues," including "open market drug dealing, assaults, robberies, public drinking, public urination, gang presence, et cetera," which made life both unpleasant and unsafe for nearby residents.[7]

Here is how Clements described the department's response:

We kept constant pressure on the corner gangs, letting them know "It is no longer your corner." We let the drug dealers know, "It's our street, time to shut down." ... Uniform officers made a ton of drug arrests as we started this approach. Every public drinker got arrested to send the message. And we continuously locked up offenders with outstanding warrants. The hookers, the junkies, the players on the corners. ... "Every time you see a known player," we told our officers, "check them in the system, and arrest them where you can." We also made a lot of car stops. The message to District 5 personnel was, "Keep the blues and reds flashing. Let them feel our presence."

Clements called this approach "zero tolerance."[8]

Probably no law enforcement strategy triggers a more hostile reaction from critics of the police than zero tolerance, with its implication of aggressive overpolicing. Many see zero tolerance as shorthand for the worst excesses of racist, militarized law enforcement, which they blame for everything from mass incarceration to the destruction of Black and Brown neighborhoods. Yet here was the chief of police apparently extolling this very strategy, proudly recounting "a ton of drug arrests" and "a lot of car stops" to "send the message." He also derided his targets as "hookers," "junkies," and "players," dehumanizing language that reduced people to caricatures. And this in a department that had supposedly shifted its focus away from paramilitary policing and become aligned with the neighborhood. A skeptic would interpret Clements's comment as proof that the department had not changed at all and was still indifferent to the well-being of neighborhood residents.

Yet it would be a serious mistake to assume that Olneyville residents do not care about behavior in their neighborhood that threatens their safety and quality of life. Many people who live in places where crime and disorder are widespread frequently implore the police to do something about the problem. These residents want law enforcement to be more engaged with their neighborhood, not less. In May 2005, when Olneyville residents met with representatives of ONE Neighborhood Builders and the police department to plan the design of Riverside Park, the residents' top goal was not to end shootings or violent crime but to eliminate the drug dealing, prostitution, and loitering in the area.[9] I once attended a community meeting in Olneyville where residents casually asked Lieutenant Richard Fernandes, the district commander at the time, about the status of a pending murder investigation but were far more interested in getting him to remove a truck that had been parked for several days under a neighborhood bridge, making it risky for people who had to cross the street. After the meeting, I asked Fernandes about this curious juxtaposition. "Oh yeah," he told me. "There could be bank robberies down the street, and that's not what we hear about. Not what they complain about. It's the little stuff."[10]

Except these things are not little to the residents of distressed neighborhoods. Parents in Olneyville and neighborhoods like it understand that most of the serious violence is confined to a very small number of young men who target one another over perceived slights or engage in risky behavior. The low-level hazards, by contrast, are far more widespread and can touch the lives of many more people.[11] During my research, I met with the executive director of one of the few nonprofits in Providence committed to radical change in the city. He lamented that when he canvassed local residents about their leading concerns, the complaint he heard most often was not about the evils of capitalism or the unequal distribution of wealth and power in American society but about homeless people sleeping in their yards. Walesca Pinto, a longtime homeowner on Appleton Street in Olneyville, told me she used to call the police "all the time" because of the sex workers plying their trade on the street outside her home. As we talked, she pointed out her window. "Right there! They were right there, all the time. Day, night. All the time!" Disorder matters.

And what I found in Providence is hardly atypical. As the Yale legal scholar and sociologist Monica Bell has observed, poor people call the police more often than wealthier people "because they depend upon police assistance in times of trouble, crises, and indecision, often to resolve noncriminal issues." In addition, "administrative data on police reporting has shown that—even controlling for crime rates—Black Americans, women, and residents of high-poverty neighborhoods are equally or more likely to call the police than other groups."[12]

So which is it? By making "a ton of arrests" and "a lot of car stops," were the Providence police part of the problem or part of the solution?

To answer this question, it helps to clarify some terms. Clements's approach was an example of "quality-of-life policing," also known as "order maintenance policing."[13] These are broad terms encompassing a number of strategies that generally target street-level crime and disorder. At its best, order maintenance policing can address the concerns most often voiced by residents in a low-income neighborhood. Yet these strategies can do great damage if implemented incorrectly. In one version of order maintenance, in an effort to reduce gun violence, a department saturates a neighborhood with officers, who stop, question, and frisk an exceedingly large number of people who appear "suspicious." The idea is that if they frisk enough people, the police will either seize guns or persuade people not to carry them.[14] (In 1968, the Supreme Court authorized the police to stop and frisk anyone "reasonably suspected" of being armed and dangerous, and courts have upheld frisks based on ambiguous cues such as "furtive gestures," evasiveness, nervousness, and action consistent with a vague "profile.")[15]

These campaigns are often called "zero tolerance" because the police stop an enormous number of people in the neighborhood, even though they know that most of the people they stop will have done nothing wrong. The problem is the police cannot reliably distinguish between the troublesome few and the innocuous many, an inability that led the architect of this strategy, the political scientist James Q. Wilson, to predict candidly that "innocent people will be stopped. Young black and Hispanic men will probably be stopped more often than older white Anglo males or women of any race." Wilson, who need not have qualified his prediction, did not consider this a serious concern.[16]

The most infamous example of this strategy is the New York stop-and-frisk campaign. Between 2002 and 2013, the NYPD recorded more than five million stops in distressed neighborhoods. As Wilson foresaw, the great majority of the people stopped were Black or Latino and had no involvement in criminal activity.[17] Though the vast majority of stops did not lead to either an arrest or a summons, NYPD officers routinely adopted the command presence typical of paramilitary policing and behaved as if they were patrolling occupied territory. In 2011, a young man surreptitiously recorded his stop and shared it with the *Nation* magazine. On tape, you can hear the police tell the man they stopped him because he glanced at them over his shoulder, an action they found suspicious. When he protested that he had done nothing wrong, they cursed him, became aggressive, and threatened to arrest him. When he asked them why he might be arrested, an officer answered, "For being a fucking mutt!" The man continued to press for an explanation, which infuriated the officers. They became increasingly aggressive, raising their voices and shouting at the man, who had done nothing wrong. As the stop unfolded, the threat of an illegal arrest gave way to the promise of extrajudicial violence. While one officer pinned the young man's arm behind his back, another threatened, "Dude, I'm gonna break your fucking arm, then I'm going to punch you in the fuckin' face." The officers eventually let him go—or rather screamed at him repeatedly to "get the fuck out of here!"[18]

Even without threats and slurs, stop-and-frisk can be intrusive and demeaning. One video, available on YouTube, shows a stop-and-frisk occurring in broad daylight in Harlem. Two young men have their palms placed against a car while two officers frisk them. The officers, who know neighborhood residents are filming the encounter, press their hands outside the two men's clothing—from ankle to crotch, along the length of each arm, and atop shirts. They squeeze buttocks, run their fingers along the inside rim of baseball caps, order the men to empty their pockets, scrutinize the contents of their wallets, and examine their identification.[19] In some cases, though not this one, residents have had to endure such an ordeal in handcuffs. Stops like these have produced seething resentment against law enforcement, an animosity not lost on the police. "The civilian population," an anonymous senior NYPD officer said, "they're being hunted by us. Instead of being protected by us, they're being hunted and we're being hated."[20] Another officer retired rather than continue to participate in the stop-and-frisk campaign, saying, "I got tired of hunting Black and Hispanic people."[21]

Nor does the damage end when the police finally break off the encounter. Scholars are beginning to pay attention to the cascading effect of *any* involvement in the criminal justice system, including a police stop. Research involving predominately Black and Latino adolescent boys has shown that routine contact with the police correlates with increases in self-reported criminal behaviors six, twelve, and eighteen months later, even independent of prior delinquency. Scholars attribute this pattern to the psychological distress produced by interactions with law enforcement, with participants who experienced more frequent police stops experiencing the greatest stress. In short, an aggressive stop-and-frisk campaign may itself generate crime.[22]

And to what end? Researchers estimate that for every 10 percent increase in the number of stops during the New York campaign, the police prevented 0.039 violent crimes and 0.131 property crimes per month.[23] The legal scholar Jeffrey Bellin has suggested that while stop-and-frisk might have had a small deterrent effect, its success depended on turning distressed minority neighborhoods into open-air prisons.[24] As for getting guns off the street, the chances that an officer would seize a gun during a stop were always low and became smaller as the number of stops increased. Between 2009 and 2012 — the peak years of the program, when the NYPD made approximately 2.4 million stops—no more than one in a thousand stops led to a conviction for either a violent crime or possession of a firearm.[25]

In another version of order maintenance, police target low-level disorder, such as graffiti, panhandling, brawling, or street-level drug dealing, to prevent a neighborhood's decline. The more familiar name for this approach is "broken windows" policing. Advocates of broken windows claim that small signs of unchecked disorder in a neighborhood open the door to more violent crime, which in turn drives away law-abiding residents and attracts people with more menacing intentions.[26] Advocates of this strategy recommend that police target the low-level behavior as soon as it appears—at the first broken window—to prevent a descent into more serious criminality. If disorder is already widespread at a location, advocates argue that the police have to act against it aggressively to send a signal that crime and disorder will no longer be tolerated there. In short, broken windows policing is all about reclaiming space. Police use it to clear an area of characters they consider undesirable so that the location can be put to better use.[27]

Like stop-and-frisk, broken windows policing is enormously controversial. For one thing, scholars disagree about whether low-level disorder, if left unaddressed, leads either to neighborhood decline or to more serious criminal behavior. The evidence is mixed.[28] But in a book that asks whether a distressed neighborhood can be transformed but not destroyed, my question is somewhat different: In the struggle to reclaim space, who stands to gain from broken windows policing? For whom is the space reclaimed? For years, scholars have found an association between broken windows policing and neoliberal strategies to lure capital back to distressed neighborhoods. Politicians and policy makers want to make cities attractive to the well-paid workers of the knowledge economy, whose high wages and free spending add so much to city coffers. These workers tend to be white, comparatively wealthy, and more credentialed than the existing Black and Latino population. When they arrive in numbers, a city also tends to see an increase in broken windows policing.[29]

The sequence has varied from place to place. In some locations, like Seattle, San Francisco, and Los Angeles, the police used order maintenance strategies to corral the homeless or other "undesirable" populations into designated zones so as to make other parts of the city more attractive to knowledge economy workers.[30] A close study of order maintenance policing in New York found that the NYPD ratcheted up broken windows policing not in gentrified neighborhoods but in the neighborhoods next to them, as if clearing the way for gentrifiers to spread to the next location.[31] Elsewhere, broken windows seems to follow or accompany gentrification, as politicians, business leaders, and new residents demand that the police either quell the "disorder" of long-term Black and Latino residents or else remove them altogether.[32] In city after city, we see a strong positive relationship between the use of neoliberal strategies to lure capital and the use of broken windows policing to control disorder.

Both stop-and-frisk and broken windows can make life much harder for the low-income residents of a distressed neighborhood. Stop-and-frisk nurses the illusion that crime can be solved solely by the police, who impose themselves on a subservient population. Worse, it treats innocent residents as part of the problem rather than part of the solution. When the police deploy an overbroad strategy like New York's stop-and-frisk campaign, they make a bad situation far worse. Because the police presence is so visible, their impact so immediate, and their authority so unyielding, people in an overpoliced community are apt to see the police as *the* problem, rather than one among many. At best, residents will see law enforcement as a problem to be solved rather than a resource to be leveraged. At worst, the police become the enemy.

Broken windows, by contrast, might be more attuned to the interests of some residents, but it may ignore the low-income residents who have endured crime and disorder for so long. By contributing to gentrification, broken windows suggests that the answer to the question at the core of this book is no, a distressed neighborhood cannot be transformed without setting the stage for its own undoing—at least not when the transformative tool is a police strategy that reclaims space for the wealthy by removing the poor. Low-income residents want the police to address crime and disorder, but not if the cure is worse than the disease.

The Providence Police Department kept up its zero tolerance strategy around the four Aleppo Street hot spots for about six months, from July to December 2003. Had the police then discontinued the strategy but made no other changes, the area would quickly have reverted to its prior state. But the zero tolerance campaign was merely one part of a collaborative effort between the police and neighborhood allies. Several of the buildings were owned by landlords who failed to comply with housing, fire, and safety codes. A task force with members from the police, the city, and the state, in close cooperation with ONE Neighborhood Builders, used these regulations to force derelict owners to bring their properties up to code. Rather than make the repairs, some owners opted to sell their property to ONE Neighborhood.[33]

ONE Neighborhood acquired the first—and worst—of the three "hot spot" buildings in December 2003, just as the order maintenance strategy ended. Acquiring this building was critical to the revitalization. It was a square, three-story, wood-frame building with two small businesses on the first floor and apartments on the second and third. Owned by an absentee landlord, the building had long been a magnet for drugs, prostitution, and brawling, including a murder in 2001. Back then, ONE Neighborhood's offices were next door, giving Frank Shea a window onto the scene. "You'd regularly see people shooting up out the window. You'd see people defecating, people dealing—just everything. I'd walk out of my office and see two junkies sticking needles in their arms."[34] And because the building sat at the corner of Manton Avenue and Aleppo Street, it threatened to choke off the revitalization of Aleppo and Riverside Park by scaring people away from the area. Over the next fifteen months, ONE Neighborhood acquired the other two properties as well as the vacant swath of land along Aleppo Street.

Once it owned the sites, ONE Neighborhood could tear down the old structures and replace them with safe, high-quality, below-market housing that, because of the shared-equity agreements, will remain affordable for low-income residents for many years. The remediation of the Riverside Mills site was then actively under way.[35] After the site was cleared, the police paired with experts in the field of crime prevention through environmental design (shortened in the literature to CPTED, pronounced "sep-ted"), who advised them on the optimal physical layout for making the park safe and inviting for residents without its becoming a haven for more unsavory uses. Finally, as part of the CPTED process, a delegation from ONE Neighborhood and the police department persuaded the city to open Aleppo Street so that it no longer came to a dead end at the edge of the park. Dead-end streets have less vehicle and pedestrian traffic, making them more isolated and increasing the risk of criminal activity. Opening the street ensured more regular use and provided more of the natural surveillance that helps guard against crime.[36] Though CPTED has been around for decades, it was new to the cops in Providence. "I'd never heard of it," Dean Isabella told me candidly.

In short, the police undertook a strategy in the area in close co-ordination with complementary activities undertaken at the same time by other actors, which collectively rebuilt Aleppo Street and created Riverside Park. Clements's approach had elements of both stop-and-frisk (it saturated an area with police in an effort to prevent violent crime) and broken windows (it targeted low-level disorder to clear an area and reclaim space). Yet it avoided the downsides of each strategy by working with neighborhood nonprofits and low-income residents, rather than apart from them, and by strictly limiting police operations both temporally and geographically. The strategy in Olneyville was designed and implemented to benefit existing low-income residents, not high-income newcomers.

I asked Dean Isabella about the department's zero tolerance strategy at Aleppo Street and Riverside Park. "The key," he told me, "is to follow the lead of the community. When it comes to strategies like this, the community gives you permission. You can't do it without permission, and you can only do as much as you have permission to do." He didn't mean formal permission, although something like that could easily be imagined. He meant that order maintenance in and around Aleppo Street, like any police strategy, had to be neighborhood approved, surgically targeted, geographically limited, and short-term—that is, only what was needed to stabilize the area being revitalized. That much, and no more, was what neighborhood residents wanted and what they permitted the police to do. Anything extra would have been excessive; anything less would have been insufficient.[37]

No one doubts the results. Between 2002 and 2007, when ONE Neighborhood completed its work in the area, reported crime at and near what had once been the worst location in the neighborhood fell by nearly 60 percent. By contrast, crime fell in the rest of the neighborhood by only 25 percent. The decline in calls for police service (reporting either shots fired, a person with a gun, drugs, or loud music and parties) fell even more precipitously. In the small area that included the four hot spots, calls for service fell by nearly 90 percent, while calls for service neighborhood-wide fell by only 40 percent.[38] These reductions in crime and disorder have endured. As the renters and homeowners who moved into the newly built housing have put down roots and begun to raise their families, Aleppo Street has matured into a stable residential area. During my research, I often ran into Olneyville newcomers who knew little of its history. To them, Aleppo Street has always been a beautiful residential address lined with well-maintained, freshly painted townhouses and single-family homes, sitting in a prime position opposite one of the nicest parks in the city. As the neighborhood memory of what used to be fades, people can hardly imagine that it was once something else.

One afternoon, I did a ride-along with a rookie Providence police officer assigned to District 5, which includes Olneyville. Fresh out of the University of Rhode Island, where he studied history, and a recent graduate of the police academy, the officer loves his job and relishes the constantly changing demands that come with policing a major city. When he asked about my research, I told him about the junkyard that used to dominate the old Riverside Mills site. "You mean Riverside Park?" "Right," I said, "and Aleppo Street used to be a mess. The police were there all the time. Prostitution. Drugs. Gangs." *"Really?"* he said. "We never get calls there now."

The relationships between the Providence Police Department and neighborhood organizations that took shape during Dean Esserman's tenure are now an accepted part of department culture. For example, the department partners with a nonprofit that provides comprehensive services to men and women involved in street prostitution. As part of an outreach program, former sex workers on the staff of the nonprofit accompany police officers and counsel men and women on the streets, educating them about addiction, treatment services, domestic violence counseling, needle exchange programs, and other matters.[39] I met with a number of these outreach workers, including a middle-aged woman with boundless energy and an engaging smile, who had been a sex worker back in the days when the Providence Police Department was still a paramilitary force. "They treated us like animals. *Animals*," she told me. "It's so much better now."

Dean Isabella is justifiably proud of the changes around Aleppo Street and Riverside Park. Standing in the park one day, near what was once the worst corner of one of the most distressed neighborhoods in Providence, he reflected on the success. "What was done, was done right."[40]

Hugh Clements designed the order maintenance strategy in Olneyville to meet the needs and respond to the demands of low-income residents. It was not imposed broadly and indiscriminately on occupied territory, nor was it implemented to make life better for wealthy newcomers. Instead it was part of a comprehensive campaign to transform a place by working in close collaboration with nonprofit organizations that were determined to make the neighborhood safe and vibrant for the long-term residents without sacrificing affordability.

Yet even here, where order maintenance succeeded about as well as can be hoped, it is important to recognize the limits of such a strategy in a neoliberal age. To begin with, neoliberalism, with its obsessive focus on wealth building by the private sector, does not pay the police to establish partnerships with the nonprofits that serve a distressed neighborhood. The same meager funding that barely supports so many nonprofits in Olneyville must now also sustain the programs that have become important to the redesigned Providence Police Department. The department has no budget for collaboration between the police and the people who counsel sex workers, for instance. The consultation with the environmental design experts who played such an important role in creating Riverside Park was arranged and funded by the Rhode Island chapter of the Local Initiatives Support Corporation, a national nonprofit dedicated to community revitalization. In short, the shift from paramilitary to neighborhood-aligned policing has no foothold in the budget of either the police department or the city. The collaborations that helped change the face of the department are no more secure than the nonprofit funding cycle. In addition, as with the creation of Riverside Park, the connections and relationships formed as a result of these collaborations principally benefit the police and nonprofits, not neighborhood residents. Residents obviously benefit indirectly from the police-nonprofit partnership, but at the end of the day, the social capital and institutional expertise reside with outsiders.

Perhaps even more importantly, when the work is complete, the very success that Clements enjoyed makes the neighborhood vastly more inviting. In a back-to-the-city moment, what ought to be an unmitigated cause for celebration becomes a source of worry. Order maintenance strategies can improve the quality of life for low-income residents, but even when such strategies are done right—as they were in Olneyville—they cannot protect a neighborhood from gentrification. Officers like Dean Isabella and Dean Esserman can align themselves with "the community" all they want. But in an age of neoliberal precarity, they cannot prevent the community's destruction.

7

"It's an Oasis"

The signs are unmistakable: the neighborhood you want to gentrify is getting better. It took a long time and a lot of hard work by a great many people, but the toxic waste dump that dragged the neighborhood down is gone and has been replaced by a magnificent park. The housing stock is much better, not just near the park but throughout the neighborhood. The worst crime hot spots are gone, and the streets are much safer. Neighborhood nonprofits have formed close, collaborative relationships both with each other and with their municipal partners. The place is almost ready to be gentrified. But one piece is missing: the local school. If you can get residents and their allies to fix the school, everything will be in place.

In June 2019, a team from Johns Hopkins University released a report on the Providence public schools. Researchers visited twelve schools—four elementary, four middle, and four high schools—where they observed classrooms, conducted interviews, and ran focus groups with students, parents, and teachers. They also met with senior administrators at district headquarters, conferred with the mayor and members of the city council, and analyzed years of standardized test scores to assess student performance over time.[1]

Their findings were horrific.

Expectations for students were chronically low. Instruction was rote and unimaginative. Students were bored and disengaged. Most children were not performing anywhere near grade level, and the longer they stayed in school, the worse it got. With only rare exceptions, teachers felt demoralized and unsupported. Parents complained that they were "marginalized" and "shut out of their children's education." Principals thought they were being held accountable for conditions beyond their power to change. Teachers and students reported that safety was "a daily concern," with "very high levels" of "bullying, demeaning, and even physical violence." Schools were collapsing, and some had deteriorated so badly that they were dangerous to all within. The worst schools "reduced seasoned members of the review team to tears."

The Johns Hopkins researchers found fault almost everywhere. The school district lacked vision and had no "clear delineations of authority, responsibility, and accountability." Too many overseers made it easy to pass the buck and hard to get things done. "There are all these chefs stirring the pot, but the soup never gets made," said one teacher. The collective bargaining agreement with the teacher's union protected incompetence, and principals (and some teachers) lamented that firing underperformers was all but impossible. Teachers complained about a chronic shortage of resources and lambasted the lack of professional development. Maintaining discipline was nearly impossible. Even the procurement process came under fire. The list of failures went on and on, but the takeaway was simple enough: The public school system was riddled with "unusually deep, systemic dysfunctions."

The report made big news. The *Providence Journal* called it "scathing."[2] A local television station described it as "devastating."[3] Even the national press took note. "Blistering," wrote Valerie Strauss, the veteran education reporter for the *Washington Post*.[4] The editorial board of the *Wall Street Journal* said the school system described by the Johns Hopkins report was "an education horror show."[5] And yet the reviewers heard stories about one school that bucked the trend. They had not visited this school, but from what they heard, it seemed like a bright light in an otherwise bleak landscape. The city had one elementary school "that offers well-funded after-school programs . . . and volunteer students and faculty from a nearby university. It's a 'full-service school with an open door to community organizations.'" They did not name this school or any others in their report, but well-informed readers knew they were describing William D'Abate Elementary in Olneyville.[6]

Brent Kerman keeps mementos—small reminders of what his students face every day. As we spoke one morning in his office at D'Abate, he reached into the top right-hand drawer of his desk, pulled out an overstuffed white envelope, and carefully peeled off the rubber band that held it closed. He opened the envelope and poured nine 9 mm shell casings and one unspent shell onto his desk. He had picked them up one morning from the playground outside the school, the brass residue of the previous night's mayhem. In one of our meetings, he showed me a photo of the houses across the street, which he took a few years ago from the grounds of D'Abate. The double-decker on the corner closest to the school was a burned-out shell. Windows were boarded shut, the foundation was crumbling, and striped traffic barrels warned pedestrians of the open pit into the basement. And that was just the house on the corner. Shoulder to shoulder along Kossuth and Florence Streets stood rows of deteriorating houses, boarded shut.

Brent has been the principal at D'Abate for eleven years, nearly three times the average tenure for public school principals nationwide.[7] For the first two or three years, the children never went outside during the school day, a decision he explained with a string of nouns that needed no verbs. "Guns, drugs, needles, pit bulls, glass." One day there was gunplay outside. An imperturbable ex-marine, Brent called the school district: "We're in lockdown. We don't need anything. I just wanted you to know." And as bad as conditions were when Brent arrived, they were even worse a decade earlier, when a guard patrolled the school parking lot full-time, and graffiti covered the building.[8]

During one of my visits, Brent asked if I had heard about the boy who was shot at Central High School at the start of the school year. When I said that I hadn't, Brent walked to the file cabinet at the rear of his office and pulled out the *Providence Journal* from September 7, 2018. The headline said it all. "Shooting Reverberates. We're Heart-broken." One student had fired at another. He missed and killed William Parsons, a former student at D'Abate. "*Great* kid," Brent told me. "Terrific family. Lives two streets away," he said, turning and pointing out his window. "His little sister is here now. In fifth grade."

Though life in Olneyville is better than when Brent arrived, chaos still surrounds the neighborhood children. He turned again and pointed to a house a few yards from the front door of the school. "You should've seen the fight going on over there. Middle of the day. Lady came out with a two-by-four, just whaling on this other woman. Mom, dad, five kids. Mom works at Subway." "Dad doesn't work?" I asked. "Ha!" he scoffed. "He works that stoop." But more than violence, it is poverty that etches the deepest scars. The vast majority of students at D'Abate are poor, and 93 percent are eligible for free or reduced-price lunches. "You see kids coming to school wearing no socks in winter. Same shirt for three weeks. No underwear. Size four shoes for a kid who needs a ten."

In this chaotic world, Brent tries "to create an oasis of calm and positivity. Kids *want* stability. We know the value of positive reinforcement. They *want* to come here." D'Abate has more than four hundred students in grades K–5; Brent knows every child by name, and most of their parents. He has the same goal for the professionals who come to work as the children who come to learn. "I try to create an environment that's professionally rewarding, and that's for everybody—teachers, staff. They feel valued, they want to come to work." Delia Rodriguez-Masjoan, a community advocate and consultant at ONE Neighborhood, told me about a snowstorm in the winter of 2019. It was a bad storm, but the city didn't cancel classes. On a day like that, you can spot the schools with a morale problem by the number of substitute teachers. At D'Abate, every teacher showed up. This may not sound like a big deal, but it sends an unmistakable signal to parents and children: The teachers care. They work hard, and the parents can count on them to be there for the kids.

I saw this on my first visit to D'Abate. I was touring the building with Brent. As we walked down the hallways, I would periodically stop to enjoy the children's artwork on the walls, and Brent would get a few steps ahead of me. Once, as I hurried to catch up, a teacher saw me scurry by her classroom. For all she could tell, I was an unknown adult walking unescorted through the school. She stopped in the middle of her lesson and marched into the hallway to confront me. "Can I help you with something?" That happened twice in one morning, as two different teachers took it upon themselves to challenge a stranger who might be up to no good. "It's an oasis," Brent told me. "A safe haven."

Brent was born in Central Falls, one of the poorest cities in Rhode Island. His father was a truck driver, and his mother worked in a jewelry factory. When he was eight, his family moved from their apartment in Central Falls to a house in North Providence, which changed everything. Rather than being raised in a dying mill town in steep decline, Brent grew up in a growing suburb with good schools, low crime, and a functioning infrastructure. "I'm so fortunate. School helped me out a lot." He joined the Marines after high school and served in Operation Desert Storm from 1990 to 1991. He began college in January 1995, graduated in 1999, and then joined the Providence school system, first as a teacher and eventually as the principal at D'Abate.

Like every great leader, Brent deflects all credit and told me time and again that the real heroes at D'Abate were the teachers. Turnover is far below the national average, and when a rare space opens up, scores of teachers apply from across the city. Brent has his pick of excellent educators, all of whom know that D'Abate is a special place to teach. Some teachers have been there for decades and have formed deep relationships with the parents and children in the neighborhood. He told me about a meeting he had with the father of a D'Abate student, which Brent recounted in his typical staccato style. "Stone cold gangster. In my office. Right where you're sitting. Like this," folding his arms across his chest and pantomiming a slouching indifference. The meeting was going nowhere until one of the teachers joined them. His visitor bolted upright in his seat. "Oh, Ms. Emily! [Not her real name]. How are you?! You remember? I'm doin' fine, yeah fine, thank you, Ms. Emily."

Much more than a public school, D'Abate is a neighborhood institution. A group of college students arrives from Brown University every afternoon to teach after-school classes. At night, the cafeteria often becomes an assembly room where residents gather to hear the mayor, chief of police, or local councilperson. Classrooms filled with children during the day hold their parents in the evening, who come after work to study English, learn how to prepare their taxes, or get tips on writing a résumé. One evening a week, a local nonprofit transforms the front entry into a food market that accepts SNAP benefits, and a local chef offers cooking classes. On Election Day, the school is a polling place. In short, D'Abate is a fixed star in the heart of the neighborhood, where residents get valuable social services, meet to air grievances, and build a sense of community. This is what the researchers from Johns Hopkins meant when they described D'Abate as "a full-service school with an open door to community organizations." Other institutions—a public library or a place of worship— could conceivably play this role, but in Olneyville, D'Abate Elementary School is the neighborhood's anchor.

But is D'Abate a *good* school?

This deceptively simple question lies at the heart of the controversy about neoliberal reform of public schools. If you do a quick Google search for William D'Abate Elementary School, you will find websites that rank public schools on the basis of widely available data. Schooldigger.com, for instance, ranks every school by its standardized test scores. In the 2018–2019 school year, about 28 percent of D'Abate students met or exceeded expectations in English and language arts, placing D'Abate 132nd out of 166 elementary schools in Rhode Island. Publicschoolreview.com compares test scores among every public school in Rhode Island, not just elementary schools. It ranks D'Abate 180th out of 282 schools. It also reports that almost all the school's students are eligible for a free or reduced-price lunch and that the student–teacher ratio is much higher than the state average. Niche.com likewise considers a few factors but places heavy emphasis on test scores and gives D'Abate an overall grade of C-.[9] On these sites, D'Abate looks like just another chronically underperforming urban school in a failing district.

But rather than trying to judge D'Abate on the basis of test scores and demographic data, we can ask what students, teachers, and parents think. Every year, the Rhode Island Department of Education surveys all three groups at every school in the state. This information is also publicly available on the department's website. Yet none of the websites that purport to rank public schools use the department's data in making their assessments.

These surveys tell a very different story. More than 85 percent of parents and teachers and 70 percent of students have a favorable impression of the overall social and learning climate at D'Abate—numbers far higher than the statewide average. Nearly eight in ten students felt the teachers set high expectations, the same as the statewide average but five points higher than other Providence schools. Six in ten students said they were invested and attentive in school, five points higher than both the state and city averages.[10] In fact, in *every* category, D'Abate scored higher than the citywide average (in one instance by more than 40 percentage points), and in nearly all it scored higher than the statewide average. Overall, D'Abate had about the same scores as the K–5 elementary school that tops the state in test scores, where 87 percent of the students are white, only 5 percent receive a free or discounted lunch, and the median household income is nearly three times that of Olneyville.[11] In short, the people who rely on D'Abate—the people for whom the school is not just a link on a website—think very highly of it. For them, D'Abate is not a failing school. It is a great success.

So is D'Abate a success or a failure?

Answering this question requires a bit of history. As critical as it was, the Johns Hopkins report on the Providence school system was not new. Similar charges have been leveled at public schools, especially urban public schools, for decades. The first and by far the most influential of these reports was *A Nation at Risk,* published in 1983.[12] Commissioned by President Reagan, it strikes a dire, some would say apocalyptic, tone.[13] "History is not kind to idlers," the authors of the report warned. Unless the nation turned its troubled schools around, it would soon be eclipsed on the world stage. "If an unfriendly foreign power had attempted to impose on America the mediocre educational performance that exists today, we might well have viewed it as an act of war." For a nation conditioned to see itself as exceptional, this was strong language.

The authors found decline everywhere. Test scores were falling, teachers were leaving, graduates could barely read. Expectations were too low, the school day too brief, and the school year too short, particularly when compared to our global competitors. The curriculum slighted the fundamentals, treating life skills such as cooking and driving on the same plane as math, history, and science. Instead of "a coherent continuum of learning," the report discovered "an often incoherent, outdated patchwork quilt." At the time, these charges were incendiary. The report dominated headlines and the national agenda, launching the modern movement for educational reform.

Other reports followed, each more urgent than the last. They began to focus more narrowly on urban public schools. In 1988, the Carnegie Foundation published *An Imperiled Generation,* which opened with the somber warning: "Without good schools, none of America's hopes can be fulfilled."[14] The authors described abysmal urban public schools across the country. A high school in Cleveland was surrounded by so many razed buildings that it "looked like a forgotten outpost in an underdeveloped country." Maintaining discipline was a constant battle, and in the most violent schools, assaults against teachers were "not uncommon." At a high school in Chicago, "only 10 percent of the entering tenth graders were able to read effectively." In the entire city of New Orleans, "the average high school senior was reading at a level exceeded by 80 percent of the students in the country." In Boston, 44 percent of students dropped out before even reaching twelfth grade. It appeared that in most cities, students were passed from one grade to the next simply to keep the pipes clear. "It's a game we play," a teacher in Houston said. "If we held them all back, the system would get clogged up. So we water down the curriculum and move them along."

Indictments of urban education began to appear with depressing regularity. In 1992, the indefatigable Fred Lippitt joined a group of parents and community leaders to form the PROBE Commission, which undertook a study of the Providence school system that was even more comprehensive than the study provided by the team from Johns Hopkins.[15] The "Providence Blueprint for Education," published in the following year, adopted a more hopeful tone than Johns Hopkins did but identified many of the same problems. Students expressed "deep frustration that many of their teachers do not respect them or take a personal interest in their lives." Teachers felt "a profound sense of professional isolation and alienation." A great majority of principals were "dissatisfied." Test scores were chronically low and fell as students advanced through grade levels. All parties passed the buck for the system's failings—from weak principals to incompetent teachers to a bullheaded union. In a word, the entire system was "cripple[d]."

From these many reports and the media coverage they inspired, a narrative has taken shape: Urban public schools are failing badly. An overstuffed bureaucracy is more concerned with enlarging itself and protecting its fiefdom than serving its students. Teachers are not accountable for their failures, and unions make it impossible to discipline or replace the worst of them. Promising students are neglected while the unruly ones wreak havoc.[16] Schools expect no better than mediocrity and advance students from one grade to the next even when they have not mastered basic skills. And because the school system has a stable supply of funding and a captive market of students, it acts like every monopoly: unresponsive to criticism, indifferent to failure, and a colossal waste of money.

Having cast the problem in these terms, education reformers argued that public schools were failing because they were not forced to compete in an open market. To set schools on the right track, reformers advised breaking up the school system's monopoly and giving parents a choice. Let them send their children to any public school, not just the failing school that happened to be nearby. But choice is irrelevant if every school in the city is burdened by bad teachers and engorged bureaucracy. To make this choice meaningful, reformers advocated creating a new kind of school, the charter school. Publicly funded but run by the private sector, charters would operate free from the traditional restrictions imposed on public schools. Teachers would not be unionized, and administrators would not be fettered by the school board. Schools could be selective, and underperforming students would not get a free pass.

In addition, all publicly funded schools, conventional and charter alike, would be forced to administer standardized tests on a regular basis. Testing promised accountability. Scores would be published so that parents would know which schools were not making the grade. Schools would have to compete for students just as businesses have to compete for customers. Parents and children could vote with their feet, and good students would no longer be stuck in an underperforming neighborhood school. Bad teachers would be removed or fired; bad schools would be closed or restructured. The stakes would be high for everyone. Failure would matter, just as it does in the private sector. In short, reformers believed the public schools needed a good old-fashioned dose of market discipline.[17]

This reform agenda did not emerge in a vacuum. It took shape as neoliberalism was gaining ascendance in American policy making, and it emulated the classic neoliberal prescription: get government out of the way and let people (in this case, parents) take responsibility for their own welfare by unleashing the private sector to meet consumer demands. Neoliberalism brought to public schools the same blind faith in the magic of the market, and the same mistrust of government, as it did to other troubled aspects of neighborhood life. *A Nation at Risk* appeared in 1983, the same year Congress created the Section 8 voucher program for housing, and *An Imperiled Generation* appeared in 1988, the year Congress made Section 8 permanent. The education reformers' solutions to the problems of public schools—choice, charter schools, and high-stakes testing—bore the distinctive stamp of neoliberal thinking.[18]

This approach to education reform has undoubted appeal, particularly for the poor. Choice is the most visible attribute of wealth in this country. Money allows people to live in a neighborhood with good schools, stable housing, functioning infrastructure, safe streets, and respectful police. A good education often provides the springboard to the financial security of the professional class in this country, while a bad one can consign a person to a lifetime of financial precarity. The very premise of this book is that the poor have the same right as the rich to live in safe, viable neighborhoods. Who would deny them a choice when it comes to their child's education? It is also true that many public schools are struggling, and many low-income students are not getting the education they need and deserve. No one who reads the Johns Hopkins report, or the many other reports written over the years, could possibly conclude otherwise.

We should therefore not be surprised that neoliberal education reforms have gained widespread support, and not only from people who have an unwavering faith in the market. The founder of the modern charter movement, Albert Shanker, was a former public school teacher and head of the United Federation of Teachers who believed charters would give teachers the freedom to develop creative, innovative schools.[19] Today some of the most committed advocates for charter schools are parents who live in distressed neighborhoods like

Olneyville, who would otherwise have to send their children to struggling or dangerous local schools.[20] In short, neoliberal education programs appealed to more than just neoliberals.

Yet the neoliberal critique of public education is deeply flawed. To begin with, its focus is far too narrow. The charges—failure to establish a rigorous curriculum and set high expectations; inability to provide a safe, orderly environment; chronically underperforming teachers who cannot be removed; and dysfunctional central administrations—all relate to conditions *inside* the school. This perpetuates the idea that a student's success in school depends entirely on what happens after she enters the building. Her performance has nothing to do with whether she arrives there hungry, cold, or frightened. If only we can fix the school, none of the rest will matter.

But we know this is not true. The numbers vary depending on the study, but the research consensus is that a student's performance in school is determined mostly by out-of-school factors, primarily the many influences associated with family and poverty.[21] Inside the school, classroom instruction and school leadership are the most important factors, but the contributions of teachers and principals pale in comparison to what goes on in other areas of a student's life.[22] Schools obviously matter a great deal, and good teachers can help narrow the gap between rich and poor students.[23] But they alone cannot close that gap. Great teachers cannot buy groceries, and dynamic principals cannot pay the rent.

Worse, by suggesting that student performance does not depend on out-of-school factors, the neoliberal critique encourages us to ignore the conditions that matter most in a child's academic achievement. As Susan Neuman, the assistant secretary of education in the George W. Bush administration, put it, "The national conversation has almost exclusively targeted schools as if they were the source of the problem, as well as the sole solution."[24] Neuman wrote these words in 2009, but the misdirection is long-standing. The authors of *A Nation at Risk* were charged with reviewing the "events in society during the past quarter century that have significantly affected educational achievement."[25] This would have been a perfect opportunity to discuss how neighborhoods like Olneyville became distressed—the history of deindustrialization, disinvestment, white flight, suburbanization, and segregation—and to describe how these seismic shifts hurt students and undermined their education. Yet the report says not a single word about this history.

Even avowedly liberal media outlets share this myopia. In the 1980s and early '90s, newspapers like the *New York Times* and professional journals like *Education Week* decried the differences between rich and poor students and demanded that the state correct the systemic conditions that placed poor students at a chronic educational disadvantage. Beginning in the late 1990s, however, concern for equality all but disappeared and was replaced by laments about an achievement gap that could supposedly be closed if only schools got better.[26] In December 2006, for instance, the *New York Times* published an editorial under the audacious headline "Why the Achievement Gap Persists." The answer, the editors wrote, was "the age-old practice" of staffing urban schools "with poorly trained and poorly educated teachers."[27] Poverty apparently played no part in the gap, and policy makers concerned with education did not need to worry about food insecurity, dangerous streets, or unsafe housing. They only needed to make public schools accountable to the ruthless efficiency of the market.

And this is the ultimate irony: it is neoliberal policy itself that has intensified the inequality and exacerbated the conditions that make it so difficult for poor children to perform in school.[28] Neoliberal reforms have decimated public support for the poor, transferred enormous wealth to the rich, and produced obscene concentrations of both wealth and poverty. For the rich, tax rates have fallen dramatically, and privatization schemes have allowed them to corral increasing shares of total wealth. For the working poor, wages have stagnated, jobs are precarious, and housing is prohibitive. Thanks to neoliberal policies, the top 1 percent of households in the United States own 40 percent of all wealth in the country, more than the bottom 90 percent combined, a higher level of inequality than at any time since at least 1962.[29] And things are getting worse. After the Trump tax cut of 2017, the four hundred richest Americans in 2018 paid a lower total tax rate, including state, federal, and local taxes, than any other income group in the country, a distinction never before equaled in U.S. history.[30] Neoliberalism produced and helps sustain the entrenched poverty that makes it so difficult for low-income students to achieve academic success, yet neoliberal solutions encourage us to ignore those conditions.

Despite the misalignment between problem and solution, the neoliberal program for education reform—choice, charters, and testing—has been widely accepted. Today about seven thousand charter schools exist nationwide, enrolling a little more than three million students, or just over 5 percent of the total K–12 enrollment in publicly funded schools.[31] Forty-five states and the District of Columbia have passed laws allowing the creation of charter schools. These laws vary from state to state, but in most jurisdictions, the legislature has freed charters from many of the rules that bind conventional public schools.[32] Roughly 5 percent of Rhode Island students have chosen to attend a charter, a bit less than the national average.[33]

Where neoliberals have had the most impact is in high-stakes testing. Properly administered, standardized tests can be a useful tool that informs both the teacher and student of gaps in a child's learning, allowing them to identify weaknesses and build on strengths. In the twentieth century, however, standardized tests became a club to use against underperforming schools. In 2001, Congress passed No Child Left Behind, the Bush administration's signature educational reform package, which sought to bring accountability to public education. It was a bet that the federal government could coerce states to provide better education for their most vulnerable children, and that this effect would be observable through steadily rising test scores.

The law demanded that all public school students become "proficient" in reading and math by 2014, with no exceptions and no excuses, though it allowed each state to develop its own measure of proficiency. Students had to be tested annually from grades 3 through 8, and once in high school. Schools that could not bring their students up to par would be subject to increasingly severe sanctions. At the extreme, bad schools could be closed, converted to charters, taken over by the state, or forced to relinquish control to private management.[34] In 2015, the Obama administration replaced No Child Left Behind with Every Student Succeeds, which returned some autonomy to the states and ended the most punitive parts of No Child Left Behind while preserving its accountability regime. Students are tested with the same frequency, and as the many websites ranking public schools reflect, test scores remain the most visible measure of a school's performance.[35]

Has the neoliberal regimen of school choice, charter schools, and high-stakes testing improved student performance and closed the divide between rich and poor students? No.

Charter schools have existed for decades, long enough to face rigorous scrutiny. The best and most thorough studies show that after controlling for demographics, charter schools perform about the same as public schools. Some charters are excellent, year in and year out, just as some public schools are excellent. Other charters consistently underperform their public counterparts. An exhaustive study in 2013 by researchers at Stanford University compared charter school students to their counterparts in traditional public schools in twenty-six states and the District of Columbia. The study found overall that white charter students tended to perform worse in both reading and math, and Black and Latino students (specifically Latinos coming from poverty) tended to perform marginally better. Asian students performed worse in math but not in reading.[36] Other studies have made similar findings. As one researcher summed it up, charter schools "have not on average performed appreciably better than regular public schools."[37]

Yet while charters offer little improvement, they can do a great deal of harm. Careful research has shown that charters can drain resources from traditional public schools. When students leave for a charter, the per-pupil spending the school district would have received goes with them. Yet the district's costs for staff, transportation, and infrastructure remain the same. The result is that charters force neighborhood schools to do the same work with less money.[38] In some parts of the country, charters are also more racially segregated than neighborhood schools.[39] And because charters can impose stricter disciplinary rules than traditional public schools, they tend to attract the more motivated learners with the most engaged parents. This undoubtedly benefits those few students, but it comes at the expense of the neighborhood school.[40]

The testing regime has produced the same anemic results as charters.[41] To begin with, testing put schools in high-poverty areas in an impossible situation by demanding that they overcome conditions that were years in the making and outside their control. And when did we decide that the whole of a child's education can be reduced to reading and math? What about history, science, and foreign language? Public schools also have ambitions—many of which cannot be quantified—besides academic achievement, like developing character, acquiring moral and ethical judgment, and preparing for responsible citizenship. The surge of standardized testing has done nothing to promote these goals.

In addition, we should be careful about equating rising test scores with improved education. Students who learn the skill of test taking can increase their scores appreciably without improving their ability either to read or to reason. The test prep industry is big business, and wealthy parents spend enormous amounts of money trying to improve their children's scores simply by making them better test takers.[42] And because teachers and schools would be blamed for their students' bad scores, No Child Left Behind baked in an obvious, unavoidable incentive: Teachers began teaching to the test. They spent precious class time making their students better test takers rather than more educated, well-rounded, and creative individuals.[43]

Even if we ask only whether high-stakes testing closed the achievement gap between rich and poor students, the answer is still no. Some test scores for some groups in some grades improved slightly under NCLB, but others declined. These improvements, however, were occurring before NCLB, and at an even faster rate. One scholar summarized the evidence: "No Child Left Behind may have had some positive effect on underserved 4th and 8th graders, but no discernible effect by the time students neared the end of elementary and secondary education. That means we have no evidence of any lasting effect."[44]

Like charter schools, however, high-stakes testing is not simply an extravagant waste of time. It also turned teachers into scapegoats, blaming them for conditions that neoliberalism helped create and teachers were powerless to change. A profession that was already underpaid is now vilified. Predictably, this has made teaching considerably less attractive, thus contributing to a severe nationwide teacher shortage. Enrollment in teacher preparation programs is down dramatically, and qualified teachers are leaving the profession at alarming rates. The magnitude of the problem varies from state to state, and a number of factors are to blame, but in surveys, teachers rank job dissatisfaction as their number one reason for leaving, and they name testing and accountability measures as the leading source of their dissatisfaction.[45] Perversely, the very measures designed to save public education are driving away the people who matter most to a child's education within school walls: teachers.

In the end, the neoliberal dream of choice, charters, and high-stakes testing has done virtually nothing to improve public education. It has battered the people and institutions central to the success of public schooling and failed to narrow the achievement gap, all while ignoring the conditions that sustain the gap in the first place. Yet the problems of public education are real, and nowhere more so than in Providence. In June 2019, Angelica Infante-Green, the Rhode Island education commissioner, began holding community forums to discuss the Johns Hopkins report and the future of the Providence school system. Fittingly, the first public meeting took place at D'Abate Elementary School.[46] Parents from across the city crowded into the cafeteria and pleaded for change. In October, supported by Providence's mayor, city council, and district school board, the state took control of the Providence public schools. The state says the take-over will last at least five years, but Infante-Green has warned that the schedule will be dictated by the pace of change, saying, "I'm not giving back the schools until it is a stable place."[47] Though parents and students widely support the move, many fear the state will impose sweeping changes without listening to the people who rely most heavily on the city's schools. Rhode Island governor Gina Raimondo insists that won't happen. "If we come up with some turnaround plan from on high and try to impose it on the community, that won't work."[48]

What will work for the Providence schools? The neoliberal program is not the solution. But if not that, then what? At last we can return to the question: Is D'Abate a success or a failure? This question has no meaning unless we also ask by what standard, and compared to what. D'Abate is a nurturing island of community and trust, a place where students, teachers, and parents feel welcome and supported. It is Olneyville's priceless gem. As for those test scores, yes, 28 percent of the students were proficient in reading in 2019. But let's drill a little deeper. As required by Every Student Succeeds, children at D'Abate are tested in grades 3, 4 and 5, and every year the percentage of students who are proficient increases dramatically. Students at D'Abate start school with weak language skills, which is hardly surprising given the number for whom English is not their first language, but the longer they stay, the better they get—both absolutely and relative to their peers statewide. By the time they leave fifth grade, about 38 percent are meeting or exceeding expectations. Only one of the twenty-one other elementary schools in Providence shows this pattern of improved progress each year, and only two have a higher percentage of students meeting or exceeding expectations at graduation. Of those three schools, none has as high a percentage of low-income students as D'Abate.

The parents, teachers, and students at D'Abate know what you cannot learn from a website. D'Abate has all the burdens that drag down countless urban schools, including many in Providence. Yet D'Abate succeeds, both as a school and as a neighborhood anchor. And D'Abate is certainly not unique. There are thousands of successful neighborhood schools, even in the most difficult environments, where students thrive despite the odds. Over time, we have learned that these schools tend to share certain characteristics. They have a nucleus of dedicated, talented teachers who feel valued and supported in their work. They have a strong, charismatic leader who supports the teachers. They have engaged parents who support the school and believe in its mission. Most of all, they work with the neighborhood rather than apart from it. They recognize that the school's fate is intimately connected with that of the neighborhood, and one cannot thrive without the other.[49]

Successful schools like D'Abate are woven into the fabric of the neighborhood. They partner with the many people and groups working to transform the area, creating a sense of community that is greater than any one person and more important than any test score. Neoliberal education reformers see neighborhoods like Olneyville as an offscreen cesspool, a place from which a small number of lucky and determined students might manage to escape, rather than a place where thousands live and deserve to thrive, where conditions are the product of official policy and therefore also the site of communal obligation. Fundamentally, neoliberal reformers see the neighborhood—and by extension the neighborhood school—as the problem. That is their great mistake. In Olneyville, the neighborhood is not the problem. The neighborhood is the solution.

The success of D'Abate Elementary School has much in common with the transformation of the Providence Police Department. In both cases, dynamic leaders forged durable bonds with community organizations that embedded the institution within the neighborhood and made it more responsive to the needs of low-income residents. D'Abate also has much in common with the construction of Riverside Park and the transformation of Aleppo Street. Jane Sherman and the Woonasquatucket River Watershed Council created a park for the entire neighborhood, not for private developers. Frank Shea and ONE Neighborhood Builders built single-family homes and apartments that will remain affordable for many years to come, protecting low-income residents from the price spirals that took place in Jamaica

Plain. Similarly, Brent Kerman and the dedicated teachers at D'Abate have created a beloved neighborhood institution built around inclusivity and community, which has allowed it to resist the siren song of charter schools. Thus, in their own way, all the changes in Olneyville succeeded because residents and their allies defied the neoliberal pull of privatization and its relentless obsession with markets that reward the wealthy few at the expense of the low-income many.

Yet for all its success, the transformation of D'Abate—like the parallel transformation of the park, the housing stock, and the police—makes the neighborhood vastly better but cannot protect it from the gentrification and displacement that improvement might herald. These changes do not merely make life much better for the people who live there and who worked so hard to produce them; they make the neighborhood increasingly attractive to the people who formerly paid it no mind.

8

"They're Pimping Us"

The transformation is nearly complete. Nearly every block of the neighborhood has been changed. The streets are safer; the school is better. Olneyville was once one of the most distressed neighborhoods in Providence. Now it may finally be ready for wealthy newcomers. But a neighborhood is more than its houses, parks, streets, and schools. What of the residents? Will they rise up to block gentrification and displacement?

To answer this question, take another look at the people who played the most visible role in Olneyville's transformation. Jane Sherman, the champion of Riverside Park, is white. So is Lisa Aurecchia, who worked with Jane for years and now continues her legacy at the Woonasquatucket River Watershed Council. Frank Shea, the former director of ONE Neighborhood Builders, is white. So is his successor, Jennifer Hawkins. In fact, whites lead nearly every nonprofit organization in the Olneyville Collaborative. In a neighborhood that is predominately Latino, whites control the organizations that have done so much to shape its destiny. None of them lives in Olneyville.[1]

Nor is Olneyville atypical. As a rule, social service nonprofits nationwide rely on well-educated, comparatively wealthy whites to identify, frame, and solve the problems endured by low-income people of color. Like any rule, this one has exceptions. But in general, whites lead and staff the organizations, just as they lead and staff the foundations and philanthropies that fund them and the boards that oversee them.[2] And because these are private organizations staffed by unelected officials, residents have no direct way to hold the organizations accountable.

Chalking the situation up to racism is too simple. Without exception, the people I met at Olneyville nonprofits are progressive reformers committed to the well-being of the low-income residents they serve. To explain the state of play, we have to look beyond individual actors to the system in which they operate. More than anything, white control of the nonprofit machinery is another unintended consequence of the neoliberal turn and the stratification and inequality it has produced. But regardless of the cause, the effect is to marginalize low-income residents in their own neighborhoods and weaken their capacity for organized resistance to the gathering threat of gentrification and displacement.

Though nonprofit organizations have existed in the United States for centuries, they did not begin to play a prominent role in distressed neighborhoods until the 1960s and did not appear in numbers until the 1970s. The timing is not coincidental. The trailblazers in this effort were community development corporations. As the architectural historian Brian Goldstein explained, the CDCs "were born of the activist spirit of the '60s—products of the War on Poverty, the civil rights movement, and reactions to the negative effects of the urban renewal program."[3] Most of the CDCs formed during the sixties and early seventies threw their roots deep into community-organized movements for local control of resources in low-income minority neighborhoods. They were consciously radical organizations whose leaders had cut their teeth on leftist politics and saw the CDC as a tool to help dismantle structural racism and systemic inequality.

At first, the federal government was more than happy to fund these organizations, though its motives changed over time. The Johnson administration saw CDCs as allied ground troops in the war on poverty; the Nixon administration viewed them as an alternative to federal involvement in cities. As Goldstein put it, "Through continued support for CDCs, Nixon could appear responsive to African-Americans while largely abdicating direct involvement in addressing urban poverty."[4] Throughout the 1970s, the federal government funded CDCs to do the work it had once done itself. This practice led to an elemental contradiction within the CDCs: they started life with a radical dream of transforming neighborhoods by dismantling government power but found themselves increasingly dependent on government support to achieve that very transformation.[5]

In the 1980s, the federal government dramatically reduced funding for CDCs even as it slashed its own support for cities. Aid to cities fell by 60 percent during the Reagan presidency, and resources intended for housing, where CDCs did most of their work, fell from almost $27 billion in 1980 to under $4 billion by 1988.[6] The abandonment of cities came just as they were beset by the AIDS epidemic, the crack epidemic, and the homelessness crisis, all of which placed enormous strains on city coffers. The demands on cities were thus particularly acute precisely when the federal government pulled most of its support.[7] In the void between a federal government that would not—and local governments that could not—support community development, CDCs gradually came to provide more and more services once supplied by state or local government.

As nonprofits replaced government in providing essential services, they also became more technocratic and professionalized. To see how this happens, consider the expertise involved in the seemingly simple task of creating a bike lane alongside an urban street, a project presently under way in Olneyville and led by the Woonasquatucket River Watershed Council. The lanes of a city street are typically about 12 feet wide, but recent research has found that drivers are safest when the lanes are 10 to 10.5 feet wide. Streets that are narrower or wider lead to more crashes, and as lanes widen, cars tend to speed up, and crashes become more severe.[8] So let's say a nonprofit, in collaboration with the city, wants to narrow the street and add a raised bike lane, leading to safer roads, fewer cars, and an improved quality of life for neighborhood residents. What does it take to get it done?

The first hurdle is always money. Depending on the size of the project, the nonprofit must cobble together funding from multiple sources, including foundations, philanthropies, government agencies, and the private sector. That's just the beginning; there are also design and engineering challenges. The nonprofit will not do the work itself but will contract with engineering companies and landscape design firms, so it has to issue a call for proposals and evaluate the bids. The nonprofit will face administrative and regulatory hurdles, since municipalities regulate city streets in meticulous detail. The police should be involved, as they were at Riverside Park, to avoid the risk that the design will inadvertently create opportunities for crime or disorder. And, of course, the nonprofit will need lawyers to write contracts and negotiate agreements, as well as to fight the inevitable battles if the project seizes or diminishes the value of private property. In short, something as apparently simple as a bike lane could demand expertise in urban design, city planning, engineering, law enforcement, municipal regulation, public finance, and law. Since no organization has all these skills in-house, getting the project done will require close working relationships with a host of professionals, as well as politicians and public servants from diverse organizations.

Because nonprofits have to devote so much time and energy to assembling the pieces of an immensely complex puzzle, they have increasingly sought out highly skilled, well-educated staff members who can bring the needed experience and connections but are often outsiders to the neighborhood. Gradually this approach severed the nonprofits from their activist roots. They abandoned any aspiration to neighborhood control and morphed into professionally managed service providers constantly on the hunt for funding. As the sociologist Jeremy Levine put it, "Paid professionals and management consultants replaced volunteers and activists; market logics replaced radical agendas. To a greater extent than ever before, a highly professionalized nonprofit sector finances and implements community development projects in poor neighborhoods."[9]

As nonprofits have become more businesslike and less radical, they have increasingly become partners in local governance. Their leadership has forged close alliances with important political actors at the state, local, and federal levels. One summer evening in June 2019, for instance, the Woonasquatucket River Watershed Council hosted a community meeting to discuss the council's plans to expand the greenway along the river. Alicia Lehrer, the council's executive director, opened the meeting by introducing Providence mayor Jorge Elorza, who praised the council and pledged to support its work. After Elorza spoke (and then promptly left the meeting), Lehrer introduced Bonnie Nickerson, the director of planning for the city, who spoke at greater length and in more detail, also in support of the council and its plans. It was obvious that Lehrer had a close professional relationship with both Elorza and Nickerson, and in an age of neoliberal precarity, such connections have become the norm.[10] By coincidence, I had met with Elorza in his office several hours before the community meeting, and when I raised my concern about gentrification and displacement in Olneyville, he acknowledged the risk but made a point of stressing the city's strong working relationship with ONE Neighborhood Builders. "We have a good partner in ONE Neighborhood," he said. "A really good partner."

These personal and professional relationships undoubtedly facilitate the work done by both the city and the nonprofits. Yet they represent an enormously important source of power that bypasses neighborhood residents almost entirely. To preserve these relationships and the political access they afford, nonprofits must prioritize collaboration over confrontation, which has led them to cast off radical activism in favor of a subdued professionalism. Nonprofits, in Olneyville and many other cities, now perform social service work, not social *change* work. They measure success by their ability to deliver results within the existing system, rather than their ability to dismantle that system and change the dynamics of power in a distressed neighborhood.

Today the nonprofits are so dominant in Olneyville that it is hard to imagine they were once otherwise. But before Jane Sherman arrived in 1993 to clean up Riverside Mills, and long before Frank Shea took the reins at ONE Neighborhood in 2000, the most important organization in Olneyville was not a nonprofit but a church. Father Raymond Tetrault is the former pastor at Saint Teresa's Church on Manton Avenue and a legendary figure in the Latino community. We met at his home on Appleton Street, where he has lived for years in a small one-bedroom apartment on the second floor of a duplex. We chatted at his kitchen table as he ate a lunch prepared for him by his landlady, who lives on the first floor. "She worries about me," he said with a smile.

Ordained in 1960, Father Tetrault began his service in the church just as Providence was experiencing its first small influx of Latino immigrants. At the time, the Catholic Church in Providence had few officials who could minister in Spanish. Father Tetrault, who speaks Spanish, recalled a man entering a store and asking where he could attend a Spanish Mass. Realizing there was no Spanish Mass celebrated in the area, Father Tetrault offered to say Mass in the man's home. Thus began his close connection with, and devotion to, the Latino community. "I learned a lot by just being with people in their houses," he said. In 1975, the Providence Diocese named him the first director of the Latin American Apostolate, and in 1990 he became the pastor at Saint Teresa's. One of his original parishioners at Saint Teresa's had been a secretary at the massive Atlantic Mills plant, not far from the church. She used to say she had been the last person in the mill before it shut down, and had locked the door behind her as she left. But that was a long time ago, and by the time Father Tetrault arrived, the mills and jobs were gone, and the neighborhood "was the pits. Half the stores on Manton Avenue were boarded up. Drugs everywhere."

Father Tetrault had been trained as an organizer, and his immediate instinct when he took over at Saint Teresa's was to organize neighborhood residents to demand change. "I always believed people should have an organizer," he told me. The goal was to use the church to give residents control over their lives. "Giving people power, letting them know that they are the church. Listening to them." I asked him what organizing meant in practice, and he answered simply: "I got up Sunday morning and said, 'We have problems. Anybody want to help?'" Father Tetrault created a group to tackle the problems identified by his parishioners, and forty people showed up at the first meeting. Every week, they met in the church basement to discuss different projects. Housing was especially important, and they got the city to lower the prices on some condemned homes so that residents could afford them. But their work varied. Ticking off some of their efforts, Father Tetrault recounted how they blocked a deportation, worked with the police to address crime and disorder, and engaged in neighborhood cleanup campaigns. Father Tetrault also had an activist's sense of the theatrical. When Jane Sherman came to Saint Teresa's to build support for the creation of Riverside Park, he arranged for a parishioner to bring buckets of water from the Woonasquatucket to remind people that the river was part of their neighborhood. It was local activism at the grassroots level, conceived and led by residents. "People had a sense of ownership, of leadership. I always thought that was very important."

In the late 1990s, Father Tetrault hired Abelardo Hernandez as a community organizer. Like many first-generation immigrants I met, Abelardo can remember the exact date he arrived in the United States. "I got to New York October 25, 1972," he told me, reciting the date as if it was his birthday. "I was fifteen." He had left his home in Guayaquil, Ecuador, and settled in Flushing, Queens, where he lived for twenty-five years. But he relocated to Manton Avenue in Olneyville with his wife and three children in 1997 to escape the crime and chaos of New York. Abelardo might be the best-known person in Olneyville, a prominence he traces to soccer. When he and his family moved to the neighborhood, his wife, Maria, began attending Holy Ghost Church in Federal Hill, where she joined the choir. There she noticed that many of the boys in the choir also played in a soccer league, but they didn't have a coach. She approached her husband, who agreed to coach the team. Maria and Abelardo started bringing lunch to the games and practices. Soon the players' families did as well, and every soccer game became an impromptu get-together of neighborhood residents. More than twenty years later, he is still the coach, and he now coaches the children of some of his first players. For Abelardo, every time they assemble is a sprawling family reunion. (To this day, Abelardo signs off on his emails, "In soccer, Abe.")

Coaching also brought him to the church. "The kids were in choir, so I said, if I'm going to work with the kids, I have to go to church too." He and his wife began attending Saint Teresa's Church in Olneyville, and Father Tetrault quickly recognized Abelardo as a leader in the neighborhood. Abelardo continued the format Father Tetrault had begun years earlier. They held regular meetings in the church basement and followed the direction set by neighborhood residents. He recounted for me some of their early campaigns. Affordable shelter remained a priority, and on one occasion a group from Olneyville descended on city hall to demand that vacant land be turned over to neighborhood residents for housing. But the community building that took place within the church was more important than any specific project. The solidarity provided residents with a connection and commitment to one another that is crucial in a distressed neighborhood. As Abelardo explained, "Low-income families suffer more than other people. The sense of community helps overcome the troubles of a low-income neighborhood. We care about one another."

Father Tetrault retired in 2008, and Saint Teresa's closed the following year.[11] Today the role the church once played in the neighborhood has been taken over by the network of nonprofits. But these organizations do not have the same connection to the neighborhood. Nearly 90 percent of Latinos identify as Christian, and approximately two-thirds consider religion "very important" in their lives and "very" or "somewhat important" in their political thinking.[12] It is no surprise that churches like Saint Teresa's are often vital in Latino community building. They have a connection to residents that nonprofits cannot easily replicate.[13]

Abelardo believes that something is lacking. The relationship isn't the same. The nonprofits hire outreach workers, but the effort strikes him as insincere, motivated only by the demands imposed by funders who want to see that residents support the nonprofit's work. "They are fulfilling the requirements to get the grants," he told me. "They go home at the end of the day. They don't live here." He mentioned a particularly prominent organization in the neighborhood as an example. "They do a good job, but they just don't have the same commitment to the neighborhood." Father Tetrault had a similar view. "The nonprofits are not so much organizers as programmers. They don't go out door knocking. It's more like advocacy programming. They do a lot of good for the neighborhood, but it's not at the grassroots."

With Saint Teresa's gone, Abelardo's role has changed. For the past decade and a half, he has been an after-school events coordinator at D'Abate Elementary School, a part-time job funded by ONE Neighborhood. He also started the local Parent-Teacher Organization. He is a constant presence at the school and has a small office tucked off a side hallway. At first blush, it might appear that ONE Neighborhood merely substituted for the church, paying Abelardo to maintain the link to the neighborhood that he had already created at Saint Teresa's. But it is not the same. Father Tetrault understood that the parishioners *were* the church, which allowed them to control its direction. Yet residents cannot *be* ONE Neighborhood, at least not in the same way. They do not own it or control it. However benevolent it might be, ONE Neighborhood is their landlord, not their church.

Dedicating a small amount of funding to support a position held by a key resident is one way a nonprofit maintains a link to the neighborhood and its residents. Another approach is to bring one or two trusted residents or local leaders onto the board of directors. For the past four years, Elmer Stanley has been the director of the Joslin Recreation Center in Olneyville, which abuts D'Abate Elementary School. Of all the people I met during my research, Elmer was probably the most insightful. He listens carefully, thinks before he speaks, and responds thoughtfully. Like Abelardo, Elmer has a history as an organizer that began with the church, and a relationship to the neighborhood and its residents that few outsiders can equal.

Elmer was born in 1958 in Norfolk, Virginia, six years before the Civil Rights Act that outlawed racial discrimination in restaurants and public accommodations, and a decade before Congress made it a crime to discriminate in housing. His mother was fifteen when he was born, and though barely more than a child herself, she instilled in her son a devotion to public service and a love of learning. When he was seven, she sent him to Boston, on his own, to be raised by her mother. "I don't want you to grow up stupid like your cousins," she said. "I hated it," Elmer told me, "but I knew it was what my mother wanted." He lived with his aunts and grandmother in Roxbury, a working-class neighborhood in steep decline, much like Olneyville. There he learned the painful lesson perennially imparted to all who are different. "It was tough to be the new kid," he said. "I got teased about my accent. Got chased home every day. Got in fights all the time." He can still remember running from gangs of kids after school, screaming for one of his aunts to open the door so that he could dart inside and escape his attackers. I asked Elmer whether he was chased by Black or white kids, and he said both, which taught him another important life lesson: a bully is a bully, regardless of skin color.

Elmer's frequent fights made him more concerned with his safety than his schooling, and the school system labeled him "a troubled child." In 1970, his mother, who by this time had also moved north, arranged for him to attend a private school outside Philadelphia. He stayed there until his last two years of high school. He was a multi-sport athlete but had not forgotten what it was like being the outsider. "I always protected the kids who were picked on, the nerds. The kids picked on because they were different." Like so many people, he credits a few great teachers with getting him to stop fighting and start studying. "Miss Mason. I still remember her name. She believed in me before I believed in myself. I took a civics class and an American history class with her. She made it come alive. I hung on every word. Same thing with an English teacher, Nancy Scartozzi." His teachers demanded the best but gave him the confidence to know he could deliver it. "I walked in scared every day, but I fell in love with Edgar Allan Poe. *The Bells. The Raven.*"

He graduated in 1977 and enrolled at Barrington College, a four-year Christian liberal arts college that has since closed. He studied social work but left school in 1982, just before graduating, married with a child on the way. After a short stint as a Pinkerton guard, he got a job with the Massachusetts Department of Corrections, first as a program coordinator for the Recreation Department ("It was a great job. I used to *kill* the guys in basketball") and later as a correctional counselor. It was good work, but he couldn't tolerate the tyranny of petty bureaucrats who could toy with a person's liberty. "Bureaucracy enrages me. We need it, but sometimes we lose the humanity in all the rules, the policies and the layers." For Elmer, prison bureaucrats were just schoolyard bullies in button-down shirts. "It was changing me. A lot of anger. A lot of drinking. I was ashamed of myself. Thank God my wife gave me time to grow up."

He left the Department of Corrections in 1990, worked as a temp for a few years at a variety of social service organizations, and eventually landed a job with a mental health organization in northern Rhode Island. In 1997, after attending a training program for community organizers, he got a job with United Interfaith Action, a faith-based organization in Fall River and New Bedford, Massachusetts, that was dedicated to serving the needs of the low-income community. He worked on a variety of campaigns, always at the direction of neighborhood residents. "Drugs. Prostitution. After school programs. What matters, regardless of the campaign, is the connection to the people. If you forget the people, it's all crap." Funded by the United States Conference of Catholic Bishops, United Interfaith Action was Father Tetrault's vision of grassroots activism at its best.

In 2001, Elmer became an organizer for the Providence Diocese.

He worked with Brazilian and African immigrants and later managed a Catholic social service agency in Central Falls, one of the poorest cities in Rhode Island. All the while, he was an active participant in distressed neighborhoods in Providence, where he and his wife have lived since 1987. He volunteered in soup kitchens and food pantries and taught Sunday school. Service comes to him as much from his faith as from his mother. "He who gives to the poor, loans to the Lord," he told me, paraphrasing Proverbs. In 2003, he met Sabina Matos through immigrant organizing, and in 2004 he managed her first campaign for city council. In 2010, Frank Shea asked Elmer to join the board of directors at ONE Neighborhood, a position he held until 2020. He told me he left because a board needs to be replenished periodically with new voices, and because he had become uncomfortable with the increasingly professionalized approach taken by the organization.

Like Abelardo, Elmer plays an important role in the neighborhood. ONE Neighborhood recognized his talent and connections when it brought him onto the board. Yet his presence on the board did not genuinely empower Olneyville residents or give them meaningful control over the organization and its direction. It is good that he was on the board, but no one could confuse his role with ownership or control of local assets.

I once attended a meeting of the Olneyville Collaborative and asked the people at the table, almost all of them young white women, whether any of them lived in Olneyville. None did. Some feared that moving to Olneyville would accelerate gentrification, which they wanted to avoid. I asked what it meant that nonprofits are staffed primarily by well-educated whites who do not live in the area or look like the people they serve. Unsurprisingly, they were familiar with the critique—it has been leveled against nonprofits for years[14]—and many had long ago noticed the same thing. But they had no response to my question.

When I asked Abelardo about the role of race in neighborhood affairs, and particularly the prevalence of white staffers and leadership in the nonprofit world, he smiled knowingly at me, as though I had unlocked a secret door. He had not brought up race before, but once I asked, he was more than happy to talk about it. His reaction was so striking that I began to ask the same question of other people of color who, like Abelardo, have achieved positions of prominence in the neighborhood. Several gave me the same knowing smile and were happy to discuss the subject. It was apparent that if I had not brought up race, they would have been content to let me languish in my ignorance. But once I asked how they felt about the fact that nonprofits play such an important role in Olneyville but are staffed and run almost entirely by whites who do not live in the neighborhood, they were eager to share their thoughts.

Elmer Stanley was characteristically thoughtful. Yes, race matters, he said, but it was important to understand *how* it matters. Elmer, who is Black, insisted that no organization actively *wants* to exclude resident voices, and staff members are universally well-intentioned. Yet they bring blind spots and biases with them wherever they go, and deploy them without knowing it. "The issue," he said, "is how do you hear the voices [of residents] and [what are] the filters you use when you hear them? Some organizations are very paternalistic." He was careful to exclude ONE Neighborhood from that critique and recalled a conversation he had with other board members about immigration and some of the challenges it creates for Olneyville residents. At first, he said, people didn't understand, but they listened respectfully and came around. Yet such understanding can only happen if someone like Elmer is in the room to offer an alternate perspective. If not, the organization remains blissfully unaware of its error.

Others were less forgiving. "They're pimping us," Iasha Hall said over dinner one evening. A longtime community activist who lives in Olneyville, she has often done community outreach for ONE Neighborhood. "I'm sorry to say it, but it's true. How many Black faces do you see at the Olneyville Collaborative? How many Brown faces? It's not empowering the community because [they] are not the face of the community. [They] are not what the people see day to day. The community is good for their power, but when it comes to listening, they're not there. They're pimping us for the 'urban experience.'" Abelardo likewise senses "a lot of resentment in the neighborhood," which he traces to the disparity between the relative wealth of whites at nonprofits and the poverty of Black and Brown residents. "There's a lot of resources in the [nonprofit] groups," he said. But though they get funding, they fail to create a genuine connection to the low-income people of color they serve. "Reaching out is one thing, but connecting and being part is something else. We live in Olneyville twenty-four hours a day."

Delia Rodriguez-Masjoan had the most nuanced take on the question of racial privilege. In early 2019, Delia became a resident engagement project manager for ONE Neighborhood, a half-time position. She is currently a consultant for the organization. Born in the United States but raised in Argentina, she returned to the United States for college and graduate school and has been an advocate for Providence's Latino community for more than two decades. Before joining the staff at ONE Neighborhood, she worked for a number of social service organizations in town. She has hosted a Spanish-language radio program for twenty-three years and has a master's degree in public administration and nonprofit management.

I asked her about the relationship between nonprofits and residents. For too long, she said, organizations have treated neighborhood residents "as an afterthought," with residents "being told things after they happen." People are too used to "they talk to us, about us, without us." I asked her whether she thought the nonprofits recognize the problem. "I think that organizations in Olneyville now know it," she wrote in an email. Unlike Abelardo and Iasha, Delia put at least part of the blame on the conditions that organizations face. "I would love to see more folks from the community working in these organizations, but [nonprofits] have a hard time finding these people, because those who work at the organization are not from these neighborhoods." Plenty of qualified residents have the experience and education, she wrote. They could do the work that needs to be done, or they could be mentored and trained to work in these positions. "But the organizations may not be ready to diversify their staff or may not be able to compensate bilingual, multicultural staff for what they are worth and what they bring to the table." Under these circumstances, residents "won't leave a state job to come to a nonprofit."

On the other hand, Delia thinks the nonprofits are not without blame. "If you say, 'I love people of color and I love diversity,' but I look at your Facebook page and it's all the people on your page who look like you, and it's not diverse, how can I trust you? How do I know what you really believe in, when your personal life is different from what you are telling our community of color?" People tend to hire people who look like them, and hiring people of color is a challenge for the nonprofits. Black and Brown residents "don't look like [the staff at the nonprofits], don't speak like them, and may have different life experiences that make it challenging to incorporate [them] into the culture of the organization." So the nonprofits hire a few "tokens of color" or invite them to be on the board but do not bring them into senior positions. Even ONE Neighborhood, probably the most important nonprofit in Olneyville, was overwhelmingly white when Rodriguez-Masjoan joined in early 2019. But she said Jennifer Hawkins, the executive director, has begun to make diversity a genuine priority. "I'll say this: ONE Neighborhood Builders has diversified incredibly over the last year and a half. Jen gets it. Jen has been a champion of it." ONE Neighborhood Builders, however, is the exception.

Abelardo's long history in Olneyville and deep involvement with the Latino community have given him rare insight into the neighborhood and the needs of the people who live there. He acquired this knowledge slowly, from the bottom up rather than the top down. After more than two decades, he has built a thick web of durable relationships throughout the neighborhood and city, giving him a credibility with residents that few outsiders can match. He doesn't have an advanced degree in city planning, urban design, or nonprofit management, but he has something no degree can bestow: the trust of neighborhood residents.

Elmer Stanley has lived and worked in distressed, low-income neighborhoods like Olneyville nearly his entire life. Social service has been his career and calling, first in the prisons, later in mental health, and finally in community organizing. His long history has taught him to stand for the outcast, heed the poor, and mistrust bureaucracy. Like Abelardo, Elmer acquired his expertise slowly, from the bottom up, through direct service and activism. He too has an elaborate network of personal and professional relationships throughout the neighborhood and city. And like Abelardo, Elmer doesn't have an advanced degree, but he has the trust and support of neighborhood residents.

Delia Rodriguez-Masjoan has been a champion of Providence's Latino community for more than two decades. She has worked for several social service organizations and, like Elmer and Abelardo, has painstakingly established a reputation as a trusted advocate.

I have not chosen these three at random. In addition to their prominence in the neighborhood and their evident skills, each has a distinct relationship with ONE Neighborhood: Abelardo's half-time position at D'Abate is funded by ONE Neighborhood; Elmer served on the board for nearly a decade; and Delia has a half-time position on the staff. Together, their positions reflect some of the ways nonprofits try to forge and maintain connections with residents. But when we look closer, we see that people like Abelardo, Elmer, and Delia play only supporting roles in the nonprofit world. They are intermediaries who link the organization to the neighborhood and its residents. Their connections to residents and community organizing, along with their intimate familiarity with the neighborhood and its challenges, lend credibility to ONE Neighborhood. But they are never the ones who control the resources that flow to these organizations. They and the low-income residents of Olneyville do not drive decisions. In a professionalized, technocratic environment, they are a small cog. Even at ONE Neighborhood, an organization trying hard to reform its practices and run by a leader who "gets it," the extent of resident control is a pale substitute for the grassroots organizing and resident empowerment that, in the words of Father Tetrault, gave residents "a sense of ownership, of leadership." In a neoliberal age, that has been lost.

And now that the nonprofits have repaired the neighborhood without empowering the residents, gentrification and displacement can finally begin.

9

Thanks for Everything (Now Get Out)

William Morgan is a Providence-based architectural critic and historian. In August 2016, he wrote a short piece for the online journal *GoLocalProv* about the Atlantic Mills complex.[1] Once the biggest employer in Olneyville, Atlantic Mills provided jobs for more than two thousand workers during its heyday in the late nineteenth century. But its machines have long since fallen silent, and today the mill is home to a few modest businesses and start-ups, including a sprawling furniture mart, a weekend flea market, and a head shop.

Though his subject was ostensibly the mill's history and architectural significance, Morgan could not resist taking a gratuitous jab at Olneyville, which he described as "pretty much a dump, especially when compared to the East Side." The East Side of Providence, home to spectacular mansions and Brown University, is the moneyed part of town. Morgan ignored the people in the neighborhood, dismissed the homes where they live ("barely hanging on" "like most things in Olneyville"), and scoffed at the businesses in the mill ("less-than-uptown enterprises").[2] And though I single out Mr. Morgan, his uncharitable view was once held by many people who live in or near Providence but spent little time in Olneyville.[3]

Yet by the time Morgan's article appeared, the perception of the neighborhood was already changing. In 2014, the *Providence Journal* recognized the "cavalry of community groups" whose efforts have been transforming the neighborhood since the mid-1990s. The paper credited many of the key steps taken by the police and nonprofits that I have described in these pages, including the creation of Riverside Park, led by Jane Sherman and the Woonasquatucket River Watershed Council, the construction of affordable housing by ONE Neighborhood Builders, the city's revitalization of William D'Abate Elementary School and adjoining Joslin Playground, and the shift in philosophy by the Providence Police Department. These organizations had provided "a critical mass of support and energy of partners and new investments that are making a big difference in the right direction."[4] Yet the paper also acknowledged that Olneyville was dogged by a "negative perception." "We have a reputation of things being worse than they are," said Providence city councilwoman Sabina Matos. "People are afraid to go there, and we have to change that."[5]

That change is happening, and a widening circle of people in and around Providence have noticed.

Perhaps nothing captures this gradual awakening better than the shifting language used to describe a single Olneyville address. The Plant is the name given to what used to be the Providence Dyeing, Bleaching and Calendering Company, a massive nineteenth-century textile treatment complex consisting of several buildings at what is now 60 Valley Street. Valley, which begins at Olneyville Square and runs north along a portion of the Woonasquatucket River, is one of the oldest streets in Olneyville and was the site of Christopher Olney's original eighteenth-century paper mill. Developers used state and federal historic tax credits to restore the Plant after the 2008 recession, retaining a small section of wall that historians believe may have been part of Olney's original mill.[6]

In 2012, when most outsiders still had a very different view of the neighborhood, a Pilates studio became a tenant at the Plant. The studio's website says it is located "on the West Side of Providence" in "a super cool neighborhood" with "plenty of parking (w/ security)."[7] In the next year, an upscale hair salon opened in the same mill complex. The website describes its space as "like a real live concrete jungle with the most amazing natural light!" and says it is located "in the Valley section of Providence."[8] The Valley neighborhood begins about ten blocks north of the Plant, past Atwells Avenue. Neither the salon nor the studio uses the word *Olneyville* on its website. The website for the company that manages the complex places 60 Valley Street "in the River District . . . next to the Olneyville neighborhood," in an "exciting, well-located mill complex."[9] Providence has no "River District," and Valley Street has been part of, not "next to," Olneyville for centuries.

But when Riffraff, a hybrid bookstore, coffee shop, and cocktail bar, opened in the Plant in December 2017, it proudly and prominently placed itself in Olneyville, only "a short walk from Olneyville Square." Though Riffraff, like the hair salon and Pilates studio, advertises the availability of free parking, it also points out the bike racks present on the property, lists the bus routes that serve the area, and feels no need to mention security. And entering the bookstore—unlike the Pilates studio—does not require that a patron be buzzed into the building.[10]

It is still possible to find writers who describe Olneyville as William Morgan did. Roadsnacks.net, for instance, described Olneyville as the second-worst neighborhood in Providence in both 2018 and 2019.[11] But the trend is unmistakably otherwise. By the time the *Providence Journal* celebrated the arrival of the "cavalry," astute observers could already read the writing on the redbrick walls of the newly restored mills. In January 2014, Providence readers of *Gawker* magazine identified the West End, a neighborhood next to Olneyville and closer to downtown, as the Williamsburg of Providence, meaning the hippest area in town, "the one with the art galleries and the boutiques and the lines for brunch." They pegged Olneyville as its Bushwick, "the Next Williamsburg."[12] In early 2017, a magazine dedicated to New England culture described Olneyville as a neighborhood "mentioned over and over again in conjunction with things that are new, interesting, and just plain cool happening in PVD."[13]

In September 2017, *East Side Monthly* ticked off the changes that best showcased how far the city had come in the past two decades. Writing in the passive voice, as though these changes had happened magically, the author described Olneyville as a place where "city parks got safer," and "old mills were converted into cool living spaces."[14] Two months later, *Providence Monthly* grandly announced "THE NEW-NEW OLNEYVILLE." Now that the city had "cracked down on safety issues and as rising rents in Providence's center have pushed people outwards," the neighborhood was expanding in "new and exciting ways," with "innovations in arts, culture, nature, and dining" that were "thrusting Olneyville into the spotlight." The piece profiled some of the artists who were making the place "a hub for visual and performing arts" and touted the trendy restaurant scene, including a spot that brought "the vision of Panamanian skate and surf meets early hip-hop and punk rock to life."[15]

Coverage like this is now the rule. Writers routinely describe Olneyville as "hip," "chic," "edgy," "artsy," or "up-and-coming."[16] For Valentine's Day 2019, *Providence Monthly* recommended a romantic dinner at Troop, "in the hip Olneyville neighborhood," and the next month celebrated the anniversary of Riffraff, "a fixture in the Olneyville renaissance."[17] A vacation website lists a "stylish studio in a historic mill" just south of Olneyville Square and says the area "is in the beginning stages of gentrification (no Starbucks). . . . Think Soho in the 80's. It is an edgy, transitional location."[18] And like Soho in the '80s, Olneyville still inspires mixed reactions from the public, as comments under these magazine articles demonstrate. While one loyal resident wrote, "Olneyville has always been the shit!," others expressed concern over the safety of the neighborhood. "Too much encroaching blight," wrote one reader, and another linked to an article about the 2018 fire on Bowdoin Street that killed Lucy Feliciano. "It disturbs me that this kind of thing happens on a regular basis in and around Olneyville."[19]

Still, positive coverage has become the overwhelming norm, and amenities in the neighborhood are now celebrated as though the recent past were a distant memory. In 2018, *Providence Monthly* ran a profile of city parks and described Joslin Playground, abutting D'Abate Elementary School, as an "exceptional stop for children, with giant boulders, native plants, rain gardens, a shade sail, a small dock and bridges, and a water park slide—not to mention softball, basketball, walking paths, and a handball court. It's also wonderful for biking," which is to say, it has come a long way from when the playground was so dangerous that Brent Kerman would not let the kids out for recess.[20]

More than just perceptions have changed. In the first decade and a half of the twenty-first century, average rents in the neighborhood climbed over 54 percent, more than twice the citywide average. Just four of the city's thirty-nine census tracts saw rents increase as fast as they did in Olneyville.[21] Between 2000 and 2018, the number of people in the neighborhood who had attended college nearly doubled, and the number with a college or advanced degree nearly tripled.[22] From 2009 to 2018, the number of non-Hispanic white residents in Olneyville increased nearly 60 percent, and the Gini inequality index, a standard measure of income inequality, increased more than 10 percent. In 2009, inequality in the neighborhood was substantially below the national rate; by 2018, it was nearly as bad.[23] In short, Olneyville in the twenty-first century has become whiter, better credentialed, significantly more unequal, and substantially

THANKS FOR EVERYTHING (NOW GET OUT)

more expensive—reliable indicators of gentrification and predictors of displacement.[24]

Chris Morrison and Jessy Sivak are the face of a changing neighborhood. They knew they wanted to leave Allston, the Boston neighborhood where they paid far too much for far too little, but other neighborhoods in Boston were even more expensive, and they had no interest in living in the suburbs. "Everyone I know lives in a city," Chris told me one afternoon. "It's like, all our parents moved to the suburbs, and all of us are moving to the cities. It's weird." They wanted a live-work space where Chris could run his business from home. Chris's uncle lives in Providence and suggested they check it out. They came for a visit, liked what they saw, and moved to the city in 2017.

At first, they lived in an apartment in the Valley neighborhood, north of Olneyville, but soon decided it didn't have what they were looking for. Chris's uncle had warned them away from Olneyville because of the crime. "Apparently some dudes kicked a guy to death in an alley around here. Not shot him or stabbed him. *Kicked* him to death. That's fucking gnarly." But they toured the Plant at 60 Valley Street and decided it was ideal. They rented a two-story unit where Chris could run his store from the ground floor and the two could live in the one-bedroom loft-style apartment upstairs. Chris was attracted to the rough-hewn spirit of the neighborhood. "It has a kind of lawlessness, you know? I like that."

Chris is the owner of the Hungry Ghost Press, an apparel and novelty store. I asked him how he came up with the name. He told me he went to college at the Jack Kerouac School of Disembodied Poetics, a unit of Naropa University in Boulder cofounded by the beat poet Allen Ginsberg. Chris dropped out after a year but studied Buddhism before he left. He told me the Hungry Ghost is one of the six realms of Chinese Buddhism, inhabited by creatures with long, narrow necks and bulging stomachs. They are insatiably hungry and devour the food that rains down around them, but because of the knots and twists in their necks, the food never reaches their stomachs, leaving them eternally unsatisfied. The Hungry Ghost thus symbolizes people who are driven by animalistic urges, destined to be forever hungry for what they need but cannot get.

"Or," Chris said at the end of his long explanation, "it's just a fucking cool sounding name."

The Hungry Ghost is impossibly eclectic. Its clothing and accessories include a collection of iron-on patches that Chris designed and created, with embroidered expressions like "Olneyville Hates You" and T-shirts that proclaim "Lonely in Olneyville." Chris sells athletic socks embroidered with the word "DEATH" and coffee mugs that say "Welcome to the pits of Hell." The aesthetic might be described as good-natured Goth. At least that's what I got from the heavy metal music on the sound system, the enormous scythe hanging on the wall behind the cash register ("That came out of my aunt's barn"), and the human skull and antique sword in the left side of the display case ("I got that on eBay"). The right side of the display case is reserved for Animal, their very friendly, bucktoothed Norwich terrier, who likes to curl up on the overstuffed pillow Chris put inside the case. Animal prefers to watch customers from inside the display case by looking at them through the mirror that lines the rear of the case.

Jessy works with autistic children, using applied behavioral analysis to help them learn what she calls "socially appropriate skills." She develops treatment plans tailored to the needs of a small group of children, working closely with their teachers, observing what works and what doesn't, and refining her plans almost daily to reinforce particular behavior. She loves her work, and though she admits it is sometimes exhausting, she finds it gratifying to help young people achieve greater independence. The two seem like an odd couple, but it works for them, Chris with his oversize earrings, scraggly beard, and eBay sword, and Jessy with her button-down sweater, advanced degree, and passion for teaching autistic children.

Chris and Jessy were eager to discuss the risk of gentrification and displacement in Olneyville, since they and many of their friends have been victims of this process elsewhere. They were forced out of Boston because rents had made it impossible to have the life they wanted. They had watched as many of their friends either moved steadily farther from the center of town or left altogether. Now that the shoe was on the other foot, it made them uncomfortable to think they were the newcomers whose arrival might force others to leave. But aside from being sensitive to the reality, they were at a loss. "What are we supposed to do?" Jessy asked. "We know that our two incomes are a lot more than what most people in Olneyville make. But we moved here because it was cheap and we couldn't afford to stay in Allston." They plan to start a family and want to raise their children in a diverse neighborhood. "That's why we're in the city. We don't want to be in some totally white suburb. We like the diversity."

Scholars and journalists have studied gentrification for decades. Some describe it as an unmitigated evil, while others see it as the salvation of distressed neighborhoods.[25] The term means different things to different people, but as Alan Mallach recently observed, "In its most widely recognized form," gentrification "is the process by which a formerly lower-income neighborhood ... draws a growing number of more affluent residents, at some point reaching a critical mass that changes the character of the neighborhood in fundamental ways." Gentrification "is most fundamentally about what happens in urban neighborhoods when they experience an influx of wealth in the form of higher-income households."[26]

Yet urban neighborhoods do not spontaneously "experience an

influx of wealth." From an economic perspective, gentrification occurs when a gap opens up between the present and potential value of land in a given area—what economists call a rent gap. When the present value of land is low but the potential value is high, capital will flood into the area to close the gap and realize the gains. If we think about gentrification simply as an economic matter, improvements in a neighborhood increase the risk of gentrification by raising the potential value of land.[27]

In thinking about the potential value of land, however, we should remember Olneyville's history. It became one of the poorest neighborhoods in Providence because capital fled and government disinvested. Olneyville did not simply wake up one morning and find itself impoverished, shorn of financial, human, and social capital, and littered with empty lots and abandoned mills. Poverty was something *done* to Olneyville. Likewise, the potential value that many now see in Olneyville did not just happen. When Jane Sherman and the Woonasquatucket River Watershed Council created Riverside Park from a toxic junkyard and transformed "Needle Park" into Donigian Park; when ONE Neighborhood Builders replaced abandoned property and vacant lots with safe homes priced at below-market rates; when administrators, faculty, and parents made D'Abate Elementary School into a community anchor; and when police changed their approach and joined with residents to establish trust and reduce crime, they unintentionally collaborated to widen the rent gap. The rent gap in Olneyville is a result of what people *did*, first to destroy the neighborhood and later to rebuild it, and the moral response to that gap must take this history into account.

In addition, when we think about gentrification, we must expand our lens beyond forced displacement. It is certainly true that many cities have become dramatically more expensive over the past two decades. It is also true that as costs have climbed, urban neighborhoods that were once affordable to people of modest means and populated largely by Blacks and Latinos have become unaffordable to all but the very wealthy and populated largely by whites. This certainly looks like displacement. Yet linking this change to gentrification is not entirely straightforward. People move for many reasons, and poor people move a lot, often with complicated motives.[28] A developer may decide, for instance, that it makes no sense to rehab a building until the police bring neighborhood disorder under control. If residents leave to avoid the harassment of order maintenance policing, they have not really been displaced by rising costs. Rather, a growing rent gap has incentivized the police to target low-income residents whose presence in a neighborhood marks it as *bad* in the eyes of the private sector. Displacement results not directly from rising costs but from the unfair alignment of police with wealthy developers and their future tenants.

Or, in a neighborhood where land values are increasing, landlords may *want* to raise rents but cannot because of rent controls. They may start to use trivial violations to justify evictions. If these tenants leave to seek out a place with a more accommodating landlord, they have not been displaced by rising costs, and yet it is hard to argue that they left voluntarily. In each case, the economics are the same (capital perceives a difference between present and potential land values and acts to close the rent gap), and the results look like displacement (poor people of color are replaced by wealthier whites). But it may be impossible to trace the departure of any specific resident to the increased costs associated with the arrival of wealthier residents.[29]

Accordingly, scholars are beginning to focus on how gentrification changes the character of a neighborhood and, with it, the lived experience of lower-income residents. In gentrifying neighborhoods, all the ratios change: rich to poor; white to Black and Brown; professional to unskilled; insured to uninsured; food secure to insecure;

holders of advanced degrees to high school graduates. Newcomers bring different preferences and the income to indulge them, leading to the establishment of different shops, restaurants, and cultural and entertainment opportunities—precisely what is taking place at 60 Valley Street—which in turn draws other like-minded newcomers. In time, the neighborhood changes. People dress differently, speak a different language, treasure different experiences, listen to different music, and buy and prepare different food. Even if no one is forced to leave, the old neighborhood no longer feels like home, leading to a condition that the British scholar Rowland Atkinson calls "symbolic displacement."[30]

In addition, decades of disinvestment and neglect have systematically drained distressed neighborhoods of social and human capital. This puts low-income residents at a disadvantage compared to many of the new arrivals, who, by virtue of their race, income, or education, carry an abundant supply of such capital wherever they go. This comparative disadvantage may be especially great in a low-income immigrant neighborhood, where residents may feel their place in society is legally and culturally precarious. They cannot leverage an advanced degree, an elaborate web of social connections, and longtime familiarity with cultural norms to navigate bureaucracies, command resources, and seize opportunities. They do not enjoy the privilege that comes with wealth and whiteness in this country. And, of course, this disadvantage is only intensified by a nonprofit regime that serves but does not empower local residents to resist the gentrification happening around them.

In short, as neighborhoods like Olneyville begin to change, low-income residents become marginalized; their political and cultural dominance in an area wanes. That is why the language used by the *Providence Monthly* to describe "the New-New Olneyville" is so important. The dramatic transformation achieved over the last two decades has given the neighborhood an undeniable cachet. But what do we learn about the neighborhood from the coverage? In Amanda Grosvenor's cover story, the *Providence Monthly* columnist assured her readers that "newcomers generally strive to respect and preserve the area's architectural landmarks and its already existent communities," but she did not explain how she knew this or why she believed it to be true. She closed her piece by stressing the need "to preserve the historic richness and not to displace the current residential community." Yet the entire article does not devote a single word to describing what that "current residential community" might be. The words "Latino," "low-income," and "immigrant" never appear. Neither does "gentrification."

Grosvenor has more in common with William Morgan, the author of the insulting broadside about Olneyville in *GoLocalProv,* than she might care to admit. Though she celebrates the neighborhood and Morgan dismisses it, both gaze on poverty without seeing the poor. Morgan fixes his sights on Olneyville but sees only "a dump." Grosvenor looks at the same place and sees the promise of radical transformation, the emergence of something "new and exciting," as though it were a tourist destination rather than an urban neighborhood. Both treat the place like a free-standing creation, floating untouched by either history or inhabitants. Humanity is all but invisible to Morgan and exists for Grosvenor only as a nondescript mass that should not be "displaced."

The *Providence Journal*'s celebration of the "cavalry" is much the same. Though the author granted that unnamed Olneyville residents also helped transform the neighborhood, only members of the "cavalry of organizations" are named and profiled in the story, and they are described as "outsiders." We learn virtually nothing about Olneyville except that it is "one of Providence's oldest and poorest neighborhoods." The residents are simply invisible and might as well be

Lithuanian as Latino. Even the article's dominant imagery offers a subtle insult: the "cavalry" in American culture classically rides in to save helpless victims from "wild Indians." Presumably we are to understand the various nonprofits as outsiders coming to rescue defenseless and faceless men and women from the predators who threaten their neighborhood. Worse, whom is the cavalry protecting? A low-income resident in Olneyville might suspect that the cavalry is there to rescue the wealthy, white newcomers, and existing residents are the menace, taking up valuable space.

To these writers, Olneyville exists without a past or a people. It is a bright, white canvas, magically restored and ready to welcome new arrivals, who can enjoy their sudden access to a hip and exciting neighborhood. To be sure, newcomers like Jessy Sivak and Chris Morrison are respectful of their neighbors and have no desire to displace them. But like most people, they travel in their own circles and choose their friends and associates from those who share their views, values, and experiences. Tolerance and respect for difference are not the same as intermingling, and experiments with mixed-income housing—built with the goal of creating casual interactions among people of different race, ethnicity, and class—have generally failed. People are tribal.[31]

Yet Chris and Jessy are exactly the sort of residents who are now attracted to Olneyville. They are young, artistic, entrepreneurial, well educated, and tolerant, with plans to set down roots and raise a family. By their presence, they add to the creative vibe and contribute to the local economy. In the city's eyes, they are ideal residents, even though their presence threatens to displace—physically, culturally, and politically—the lower-income Latino residents who live nearby. Delia Rodriguez-Masjoan, the resident engagement manager for ONE Neighborhood and a longtime advocate for the Latino community, shares my concern about Olneyville's future and worries about gentrification and displacement. Twenty years ago, she worked at a different social service organization in Olneyville. Back then, "The neighborhood was all the things associated with 'negative' in the English language." Now, recognizing that people outside Olneyville have suddenly become interested in the neighborhood, she is uneasy about the changes taking place. "'Let's put in this new restaurant, these coffee shops.' That's not transformation *for the neighborhood*," she says. "It's not gonna be the same neighborhood."

In late 2018, Providence released the Woonasquatucket Vision Plan, an ambitious proposal to revitalize an arc of formerly industrial land on either side of the river. The arc starts at Olneyville Square, wends north along Valley Street, and gradually bends east as it passes through Olneyville and two other neighborhoods. Building on trends already under way, the city hopes to transform this land into an "innovation district" that will serve as a mecca for artists and creators, drawing more newcomers like Chris and Jessy. A response to the rise of the knowledge economy, innovation districts are compact areas in a city where institutions like hospitals and universities cluster near like-minded firms, start-ups, and business incubators that provide access to mentors and investors. Innovation districts are technologically wired, accessible by mass transit, friendly to walkers and bikers, and served by an entire ecology of restaurants, bars, and niche businesses. They are frequently developed in or near distressed neighborhoods, where land remains relatively cheap. Today more than one hundred innovation districts are in various stages of development around the world.

The most successful of these districts leverage an area's unique physical, cultural, and economic assets to create a dynamic, open, and competitive energy that drives innovation and an entrepreneurial spirit. Along the Woonasquatucket, the city wants to develop the unused land and mill buildings to create sustainably affordable space for artists and creators. Providence rightly prides itself as a haven for the visual and performing arts, and Olneyville has long been the center of an off-the-grid art scene.[32] The city hopes to create a distinctive space, with a cultural identity built around artistic creativity that supports the art scene while incubating and developing high-paying knowledge-based jobs. This atmosphere will then draw the well-educated, well-paid demographic that has transformed so many urban neighborhoods. As the authors of the vision plan observed, innovation districts "are being used throughout the globe as this century's 'productive geography' much like the historic mill structures found throughout the Woonasquatucket River valley are remnants of the last century's productive geography. It is fitting that these structures now be repurposed to further economic, cultural, and social pursuits."[33]

That is what Providence wants, because it is the only strategy that will fill the city's coffers after forty-five years of neoliberal austerity. In fact, the city has already made precisely this bet in another part of town. In late 2012, Providence released its plan for the creation of what it calls the Knowledge District, an attempt to leverage its leading anchor institutions—including Brown University and Rhode Island Hospital—to create a concentration of high-tech, high-paying, knowledge-based jobs in the old jewelry district abutting downtown (or Downcity, as people call it in Providence).[34] Years in the making, the Knowledge District has attracted Brown's new medical school and associated laboratories, as well as the toymaker Hasbro.[35] The innovation district along the Woonasquatucket would follow the same path that the city committed to years earlier.

It is far too soon to tell whether the city's plan for the innovation district will succeed. In December 2019, Gotham Greens opened a 100,000-square-foot hydroponic lettuce farm in the district on the site of an abandoned General Electric plant. The urban farm, which will grow six million heads of lettuce annually for distribution through-out New England, is a quintessential twenty-first-century high-tech enterprise. Powered entirely by renewable electricity, the farm uses 95 percent less water and 97 percent less land than conventional let-tuce farms. Yet it will have only sixty employees, a far cry from the thousands who used to work in the mills, and it is unclear how many will come from Olneyville.[36] Apart from the farm, however, much of the remaining land reserved for the innovation district, especially within Olneyville, remains pocked with abandoned mills and vacant lots. And no one should underestimate the magnitude of the chal-lenge. One of the industries lured to the Knowledge District, a video game company started by former Major League Baseball pitcher Curt Schilling, collapsed spectacularly and defaulted on a $75 million loan guaranteed by the State of Rhode Island, leaving Rhode Island tax-payers to foot the bill.[37]

Even if the innovation district attracts more employers like Gotham Greens, the plan presents immediate problems that threaten to overwhelm the neighborhood. The first is what the city calls the "inherent conflict" between creative artists and the wealthier demo-graphic of the knowledge economy. The former cannot survive with-out sustainably affordable space to explore, experiment, and cre-ate; the latter, with its greater disposable income and consumerist tastes, drives up the cost of land and threatens to displace the art-ists for whom the district was created. That's what happened in Sea-port, Boston's innovation district, which saw rents for office space jump 43 percent in its early years—making it exceedingly difficult for the small businesses that initially fueled the district's growth to con-tinue operating there. Seaport is now home to corporate giants like PricewaterhouseCoopers, Goodwin Procter, and State Street Corpo-ration—not exactly the artistic start-ups and entrepreneurial incu-bators that innovation districts aim to attract.[38] A second conflict over land also looms on the horizon: in time, the upward pressure on rents may drive early arrivals to the innovation district farther west, across the Woonasquatucket River, into Olneyville's Latino residential core. It is not clear that these conflicts can be avoided.

In addition, while a strong risk exists that an innovation district will drive up rents, there is little likelihood that it will produce good jobs for poor residents of the neighborhood (as opposed to white, comparatively wealthy newcomers). While the fastest-growing U.S. cities are adding jobs, a number of studies have shown that this job growth mainly benefits people who are well-off. As befits its name, the knowledge economy rewards education and technological skills. Those who can leverage their expertise are handsomely rewarded, but those who cannot are typically relegated to low-wage jobs in the service sector, amplifying the inequity of the hourglass economy. These new jobs also exacerbate racial inequality. Investments and job growth in major cities tend to benefit whites rather than Blacks.[39] Nothing in the nature of an innovation district avoids this problem, and early experience with other districts is not auspicious.[40]

In short, the city is pinning its hopes on the knowledge economy to generate high-paying jobs with no plan in place to improve the well-being of the people whose labor made the neighborhood so desirable. Many more people like Chris and Jessy will move in, and we can fairly predict that most will be exceedingly tolerant toward, but disengaged from, their Latino neighbors, who will receive little benefit from the new wealth. The newcomers will almost certainly have more social and human capital than their low-income neighbors, be much better able to define and collectively defend their interests, and will contribute far more to the city's bottom line. Scholars have long shown that public policy tends to favor the preferences of the haves over the have-nots, so if disagreements arise between the rich and poor residents of Olneyville, the wealthy will probably get their way.[41] Unless controlled, therefore, the innovation district will likely continue the trends already under way in the neighborhood. Even if the city's administrators recognize the risk and try to avoid it, the economic and political forces operating in today's neoliberal environment all but guarantee that the neighborhood will continue to get wealthier, whiter, less equal, and more expensive.[42]

Signs indicate that these trends will not only continue but accelerate. In 2017, Congress passed and the president signed the Tax Cuts and Jobs Act. Among other provisions, the law allows the governor of each state to designate a number of low-income, high-poverty census tracts as Opportunity Zones. Private investors can reduce or eliminate taxes on capital gains if they invest in real estate or businesses in these areas. The size of the tax break depends on how long the taxpayer holds the investment. If she maintains it for at least ten years, she can sell without paying any capital gains tax at all. In late 2018, Rhode Island governor Gina Raimondo designated Olneyville an Opportunity Zone.[43]

Opportunity Zones are meant to draw private capital into distressed areas in the hope that investment will stimulate economic growth. Laws that encourage private investment for the ostensible benefit of the poor are not new. On the contrary, incentives of this sort are the bread and butter of neoliberal policy and have been championed by Republicans and Democrats alike. The Low-Income Housing Tax Credit, for instance, gives developers a tax incentive to finance the construction of quality housing priced at below-market rates.

But unlike other tax incentives, the Tax Cuts and Jobs Act leaves the capital lured into Opportunity Zones remarkably unrestrained. The law creating Opportunity Zones mandates investment in a particular place and obligates investors to make improvements to the properties they purchase. It also excludes certain "sin" businesses from the subsidy, such as massage parlors or gambling houses. Otherwise, however, investment is completely unregulated. Though the law vaguely implies that investments in Opportunity Zones will benefit low-income residents, this provision is at best unenforceable, and at worst a deliberate bait and switch. As the National Low Income Housing Coalition advised its members, "There are no regulatory provisions specifying that investments must benefit low-income people, build affordable housing, employ low-income residents, or protect and support existing local business. Nor are there protections to prevent displacement of low-income people as a result of the new investments in the distressed communities." NLIHC president and CEO Diane Yentel correctly observed that unless the government establishes "regulatory guardrails" to restrain capital, "there is no guarantee that low-income people will benefit in any significant way—if at all—from Opportunity Zones."[44]

In the absence of such guardrails, billions of dollars of untaxed profits have been poured into projects that offer little or no conceivable benefits for low-income residents of Opportunity Zones. In Houston, the tax benefit is helping to build a glass-wrapped skyscraper that will contain luxury apartments, a yoga lawn, and a pool surrounded by cabanas. In Miami, it is being used to finance an upscale office tower with a landscaped roof terrace. Both of these structures are in neighborhoods that are already seeing skyrocketing prices and an influx of projects aimed at the wealthy. Also notable is that many of the projects in Opportunity Zones were under way before the law passed, meaning developers had already concluded that the projects warranted the investment. Now they will reap greater profits for a risk they were already prepared to take.[45]

The investments in Houston and Miami are not a coincidence; the law is actually structured to increase the likelihood that capital will flood into areas on the cusp of gentrification. Adam Looney is an expert in federal tax policy. As he explained in a research memo for the Brookings Institution, the law works its magic by promising to reduce or eliminate the tax on capital gains. The value of the tax benefit, therefore, will be greatest in the areas where the investment grows the most. This in turn will draw investor dollars to places where land values are low but rising rapidly, allowing investors to buy low but sell high, and to businesses that attract consumers with the highest disposable income. In short, Opportunity Zones will be most profitable in areas that are gentrifying, like Olneyville.[46]

Lawmakers are not blind to these concerns. Some Democrats who initially supported Opportunity Zones — including some members of the Congressional Black Caucus who represent the neighborhoods that Opportunity Zones were ostensibly designed to help — have drafted corrective legislation. One of their bills would eliminate some two hundred Opportunity Zones in relatively wealthy districts, and another would require investors to file annual reports detailing their development projects and compel the Treasury Department to publish reports on job creation and poverty reduction attributable to Opportunity Zones. These changes would at least increase transparency and limit the windfalls going to wealthy investors, though they do nothing to guarantee greater benefit for distressed neighborhoods. In any case, as of spring 2021 none of this legislation had become law.[47]

It is possible that some investors in Opportunity Zones will be motivated to benefit low-income residents rather than maximize their own returns. In at least some Opportunity Zones, investors appear to be taking this approach. The Pennsylvania Flagship Opportunity Zone in Erie, for instance, has attracted investments in tech labs for public schools, transportation improvements, broadband deployment, and a reentry program for people who were formerly incarcerated.[48] But nothing in the law commands such benevolence, and nothing in Olneyville's history gives us reason to believe altruism will be the rule.

In any case, Rhode Island and Providence do not seem to be targeting altruists. We know relatively little about how the city and state chose their Opportunity Zones, since they have not disclosed their criteria or decision-making process. The state website for the program, however, reports that the Rhode Island Department of Commerce asked local municipalities to identify eligible census tracts where private investment would be "most viable." Though they also sought to identify tracts where that private investment "would produce the greatest community benefit, and would be supported by local economic development efforts," the first condition was the creation of private wealth. The entire website seems focused on the potential benefits to wealthy investors.[49]

We have yet another cause for concern. Because the benefit to investors in Opportunity Zones primarily comes in the form of a smaller individual tax bill, and because personal tax returns are not public records, we have no way to know how many people are investing, what investments they are making, what businesses or property they intend to develop, or which consumers they hope to reach. Bonnie Nickerson is the director of the Providence Department of Planning and Development. During one of our interviews, she told me about a meeting she had with private investors who wanted to buy the old Paragon Mills complex along Manton Avenue in Olneyville. Nickerson cut the meeting short when she learned that the investors planned to create a gated upscale residential community, entirely out of character with the neighborhood.[50]

Nickerson can exercise that power only because Providence owns Paragon Mills. Most of Olneyville's mills are privately owned. I asked her whether anything would prevent the developers from taking their exclusionist vision to a different mill in Olneyville in private hands, and she shrugged. "If we don't own it, there's nothing I can do." In short, Opportunity Zones raise the specter of capitalism unleashed:

invisible investors making unknown investments for unknowable reasons. The absence of any leash is what led one urban planner to lament that the law encourages "gentrification on steroids."[51]

If Olneyville has anything in its favor, it is that gentrification is still in the earliest stages. Though whites are arriving in numbers, the neighborhood remains predominately Latino. Moreover, when developers rehab a mill, they add to the available housing stock, which allows the neighborhood to grow without physically displacing existing residents. In addition, though rents in Olneyville are increasing substantially faster than in the rest of the city, they remain well below city averages. Olneyville is still a poor neighborhood, and none of the residents I spoke with felt they were currently threatened by physical or cultural displacement (though many complained about rising costs, and a few spoke of businesses and restaurants that closed because they could not afford the rent increases imposed by their landlords).

Still, nothing about the state of play should reassure us about Olneyville's future as a sustainable, affordable neighborhood. As more and more newcomers arrive, the poor will begin to feel the symbolic displacement that augurs a changing neighborhood. Physical displacement tends to follow soon after; when it begins, neighborhoods can change quickly. One study of the nation's fastest-gentrifying zip codes showed that in downtown Los Angeles, median home value increased by 707 percent, and the population holding a bachelor's degree or higher increased by 857 percent, between 2000 and 2016.[52] Washington, D.C., which ranked second on the same list, saw more than twenty thousand Black residents displaced from their neighborhoods between 2000 and 2013.[53] Even in other parts of Providence, small homes like the ones that line the residential streets of Olneyville have been rehabbed and fetch prices that would have been inconceivable only a few years ago. In Olneyville, this transformation is well under way. Between 2012 and 2018, the average value of real estate sales in the neighborhood rose from approximately $287,000 to $393,000.[54] We cannot know how much time Olneyville has, but if experience elsewhere holds any lesson, it is that once the process reaches a tipping point, it cannot be reversed.

In 2014, councilwoman Sabina Matos lamented that people were afraid to come to Olneyville, and vowed to change that perception. Thanks to years of hard work by the "cavalry," she has succeeded. But to what end? In 1989, when Riverside Mills burned to the ground, no one foresaw the new Olneyville. Throughout the 1990s, as Providence began to bask in the national spotlight of a "renaissance," Olneyville sank deeper into blight. When city and state fathers poured billions of dollars into downtown but left the charred, toxic carcass of Riverside Mills where it lay; when children at D'Abate Elementary School had to stay inside for recess because their overgrown playground was littered with used syringes and a guard patrolled the school parking lot; when the Providence Police Department had one of the highest rates of police brutality in the nation, most people had given up on Olneyville.

This is the place people like Jane Sherman, Frank Shea, and Dean Esserman set out to change. They have been remarkably successful over the past twenty-five years. Yet what will come of their energy and talents? They helped bring the neighborhood to the attention of an entirely new category of residents. These newcomers want the same thing as the people who have lived there for decades: a safe, affordable, thriving neighborhood. But like newcomers everywhere, these men and women are creating a different environment that suits their tastes and income. And as these newcomers began to arrive, Olneyville became a perfect candidate for a massive federal subsidy that gives wealthy taxpayers an incentive to accelerate a transition toward white wealth that was already well under way. In our neoliberal age, the success of this transition is measured by what benefits wealthy investors rather than poor residents—by the metrics of the market rather than of the neighborhood. It therefore threatens to leave Olneyville at the mercy of unrestrained capital.

The world of neoliberal precarity transformed Olneyville but cannot protect it.

10

Trust the Neighborhood

Having traced the arc from "Lonelyville" to "the next Williamsburg," we can finally put our finger on what is wrong with the current approach to neighborhood well-being. When it works—and in Olneyville it has worked about as well as can be hoped—it brings in vast resources to restore physical space and rebuild infrastructure, but those resources are not controlled by the neighborhood and its residents. The current approach can establish strong and durable relationships with powerful local, state, and national politicians and foundations, but it is the executive director of a nonprofit, not neighborhood residents, who has the mayor's chief of staff on speed dial. The current approach can establish programs that target a range of local needs, but outsiders will judge what programs should be funded, how they have performed, and whether they should be continued. The current approach can build a smattering of houses and apartments at below-market rates, but there will never be enough homes to shelter the neighborhood's most vulnerable residents. The current approach can improve life for the people who call the neighborhood home, but it cannot empower them. In the end, the current approach can transform a neighborhood, making it attractive to people with greater wealth, but cannot protect it from the destruction that may follow.

Low-income residents in neighborhoods like Olneyville need a new tool kit that gives them the protection, security, and social capital that the wealthy enjoy simply by virtue of their wealth. Low-income residents need a stream of funding that is adequate and secure, but not controlled by outsiders. They need connections to, and relationships with, people in power. They need a mechanism that decommodifies property so that capital cannot displace longtime residents and businesses. They need a tool that allows the people in the neighborhood—rather than an outside organization—to capture increases in the value of land. And they need weapons to help ensure that as capital approaches, public institutions like schools and the police are aligned with the poor rather than the wealthy.

What they need is an innovation that I call a neighborhood trust.

A trust is simply a legal instrument that owns assets for the benefit of some other individual or group. The group could be a family or an organization, but it can also be the residents of a particular place, like a neighborhood. The assets can be anything of value, including land, buildings, cash, businesses, and other things in a neighborhood. These assets are overseen by a group of trustees, who manage and administer the trust pursuant to a set of rules that are established when the trust is created, and can be modified over time to meet the changing needs of the group. The trustees do not own the wealth; the trust owns the wealth for the group's benefit. But the trustees oversee it according to rules designed to protect, preserve, and increase that wealth. Trustees are chosen or designated when the trust is created, and they can serve indefinitely or for a fixed term.

A couple of examples will help clarify how a trust works. I live in Ithaca, a small town in upstate New York and the home of Cornell University, where I teach. Ithaca sits at the southern tip of Cayuga Lake, one of eleven Finger Lakes. Rich farmland rises steeply from the shores of these lakes and rolls across the horizon, creating a landscape of undulating hills, deep gorges, and thundering waterfalls. It is a place of stunning natural beauty, which many people want to preserve for future generations. In 1989, a group came together to create the

Finger Lakes Land Trust, whose mission is "to conserve forever the lands and waters of the Finger Lakes region, ensuring scenic vistas, clean water, local foods, and wild places for everyone."[1]

The trust fulfills its mission by acquiring land and creating a chain of nature preserves that are open to the public year-round. The trust owns the land and holds it for the benefit of the public. A board of directors oversees the trust, though a paid staff of professional managers and conservationists manages its day-to-day operation. People can contribute to the trust by donating land, either as a gift or as a bequest, or by contributing money or other assets, which the trustees can use to purchase additional land. To date, the trust has brought nearly twenty-four thousand acres under its protection. This land cannot be purchased on the private market. No matter how much private investors offer, the trustees are not permitted to sell the land.[2]

The urban counterpart to a conservation land trust is a community land trust. While the former acquires and decommodifies rural farmland and forests, the latter acquires and decommodifies urban property. Both hold the assets for the long-term benefit of another group—in the former case, the public; in the latter, the residents of a particular area. Both are overseen by a board of trustees, who follow a set of rules established at the creation of the trust, and both are managed by a professional staff. Community land trusts have existed for decades, and hundreds of successful models exist all across the country.[3]

The oldest community land trust, created in 1969, is in Georgia; the largest is in Burlington, Vermont.[4] The most famous, however, and the one whose operation provides a model for a distressed neighborhood like Olneyville, might be the Dudley Neighbors Incorporated land trust, in an area known as the Dudley Triangle in the Roxbury and Dorchester neighborhoods of Boston. As in Olneyville, the economic downturn of the 1970s and '80s devastated the triangle. Vacant lots became toxic dumps, and abandoned buildings burned to the ground. By 1984, after years of disinvestment and arson, nearly one-third of the triangle's land lay vacant. In the same year, a group of determined residents formed a nonprofit community group, later named the Dudley Street Neighborhood Initiative, to organize and mobilize area residents against the illegal dumping and arson that beset the area.

When the City of Boston circulated a neighborhood redevelopment plan that included no input from Dudley residents, DSNI responded. Area residents researched and drafted a comprehensive alternative to the city plan. Under the rallying cry "Take a Stand. Own the Land!" residents settled on a community land trust as the best way to decommodify land to preserve its long-term affordability. To implement its plan, DSNI created Dudley Neighbors Incorporated, a community land trust. In 1988, Boston gave the land trust the power of eminent domain, which allowed the trust to acquire much of the privately owned but abandoned or vacant land within the triangle and incorporate it into the trust. The city also owned a great deal of vacant land in the triangle, which it transferred to the trust. Today the trust owns the land and buildings on this land for the benefit of the neighborhood's residents.[5]

Within the trust, DNI gradually developed the land as permanently affordable housing by decoupling it from the unrestrained forces of the market. Under an arrangement much like what ONE Neighborhood Builders uses in Olneyville, buyers who purchase a home built by DNI own the structure but not the land beneath it. This creates the classic quid pro quo of shared equity: because the homeowner does not own the land, she can buy the house at a considerably lower price. In exchange, she agrees to give up the right to sell for whatever price the market could demand. Instead, if she chooses to sell, she must sell to another eligible low-income buyer and cannot sell at the market rate.

As in Olneyville, these shared-equity arrangements have helped the Dudley neighborhood thrive by ensuring a stable supply of quality homes at below-market prices. They enable homeowners to resist displacement and give them a greater attachment to the neighborhood by reducing their financial incentive to sell. They also eliminate the risk that speculators with no commitment to the neighborhood will snatch up distressed homes and flip them at prices beyond the reach of the average neighborhood resident. Though land trusts typically focus on homeownership, their essential benefit is to decommodify land, which makes them equally appropriate for rental units. Most people living on land owned by a community land trust, in Dudley and across the country, rent rather than own. They live in multiunit buildings owned by the trusts, which rent units to residents at affordable rates. Land trusts can also decommodify commercial property, helping to ensure that businesses serving the needs of low-income residents will not be displaced by rising rents. Though DNI originally focused on affordable housing, it has gradually expanded. As of 2016, the trust managed 226 homes, an urban farm, a community greenhouse, a charter school, several parks, and a town common.[6]

Yet there is a critical difference between the shared-equity arrangement created by ONE Neighborhood and the arrangement created by DNI. In Olneyville, ONE Neighborhood owns the land; in the Dudley Triangle, *the residents of the neighborhood collectively own the land*. In Olneyville, the residents in shared-equity housing divide their wealth with ONE Neighborhood, an outside nonprofit; in the Dudley Triangle, they divide it with the rest of their neighbors. Increases in the land's value accrue to the neighborhood as a whole rather than to an outside steward, however benevolent. Jennifer Hawkins, the current executive director of ONE Neighborhood Builders and a passionate, dedicated advocate for Olneyville, understands this distinction. I once attended a meeting with Hawkins and other members of her staff and raised the idea of a community land trust. She nodded. "I don't *want* to own Olneyville," she said. She was speaking for the organization, of course, not for herself. Yet despite this preference, ONE Neighborhood remains one of the largest private landowners in Olneyville. It owns some of the most valuable residential space in the neighborhood. Increases in the value of the land accrue to ONE Neighborhood, not to residents, and ONE Neighborhood decides how the land will be developed and used. The welfare of the neighborhood continues to depend on the magnanimity of outsiders. This is neoliberal precarity in action.

The community land trust is an exceedingly important model for neighborhood well-being, and CLTs have received a great deal of attention as a bulwark against gentrification and displacement. If this book does nothing more than contribute to the groundswell of support for CLTs, I will consider it a success. Yet the community land trust does not go far enough toward creating vibrant, sustainable low-income neighborhoods. To begin with, most land trusts are far too small to stabilize an entire neighborhood. They provide long-term affordable housing to the people living in houses and apartments owned by the trust, but that is not enough to protect the neighborhood from gentrification and displacement. Moreover, though land trusts decommodify property, the interconnected needs of a distressed neighborhood go well beyond preserving the affordability of land. In a neoliberal age, the government will not—and the private sector cannot—meet these interconnected needs. To fill the gap, neighborhoods like Olneyville must rely on the ecosystem of small nonprofit organizations, which run desperately important programs and services but operate on a shoestring. If these organizations close, the neighborhood becomes practically unlivable for the people who depend on them. Preserving these organizations, and bringing them within the stable ownership and control of neighborhood residents, is just as important as decommodifying land.

Many community land trusts have also set their sights too low. As the authors of one study found, the staff and leadership of CLTs often "do not challenge the larger relations, processes, or institutions of society." Instead they see themselves as making it easier for low-income residents to survive in a neoliberal state that has left them behind. Like nonprofits in Olneyville and elsewhere, many CLTs are service providers rather than neighborhood activists.[7] As laudable as that may be, it is not enough. As Father Tetrault understood many years ago, neighborhood residents must *own and control their fate.* Neighborhoods like Olneyville need a model that replenishes social capital, restores a sense of community, and gives its residents ownership and control of local resources, including but not limited to the land. That is the motivation behind the neighborhood trust.

Suppose we established a fund, held in trust by Olneyville residents for the long-term benefit of the neighborhood and its low-income residents. Let's call it the Olneyville Neighborhood Trust. All the resources that flow into the neighborhood from philanthropies, foundations, community banks, and governments, which now go to nonprofits, would instead pass into and be administered by the trust. The trust would own and control these assets and would have a mandate to address the interconnected challenges that confront the neighborhood, but in a way that preserved sustainable affordability. Beyond that broad mandate, the residents could tailor the trust's terms and objectives to meet the unique needs of the neighborhood. As those needs changed, so could the terms of the trust, but only if the residents of the neighborhood agreed. A board of trustees would administer the trust, and a majority of the board would always come from the neighborhood and its trusted allies: the residents, business owners, and civic leaders who best understand the place and what it needs. They would be elected for rotating terms of office by their neighbors, giving the residents a direct voice in the direction of the trust and the disposition of its assets. This would restore the democratic accountability that is absent in the current model, and make the trust an engine not simply of sustainable economic and neighborhood development but also of civic participation. As in any trust, the board could hire a professional staff to manage daily operations.

In time, a neighborhood trust would fill the gaps opened by neoliberalism and end the precarity it has created. To begin with, the trust could decommodify land by creating shared-equity arrangements in commercial and residential property, just as community land trusts do now. But a neighborhood trust would go substantially beyond this. It would give residents local ownership and control of the resources that flow into their neighborhood, liberating them from the stewardship of outside nonprofits. Not only is this important in its own right, but maintaining ownership and control encourages neighborhoods to develop and tap local residents' expertise, thus building essential skills in self-governance and ending the dependency that too often defines the current model. That was the core insight of Father Tetrault's work at Saint Teresa's: placing power in the hands of the people affected by neighborhood problems builds leadership and self-sufficiency. Once the trust controlled these resources, residents could decide how they should be deployed. If, for instance, the people living in Olneyville believed that English for Action provided a more valuable service than some other organization in the neighborhood, they should be allowed to allocate their resources accordingly. That choice should not be made by an outside foundation. Organizations would thus be accountable to the neighborhood and the people who live there, rather than to a distant philanthropy. The neighborhood would be better off, since the needs and wants of low-income residents would be catered to directly.

In addition, a neighborhood trust could invest its resources to create an endowment—a trust fund—that would help it weather economic downturns, survive the inevitable shifts in priorities by funders, and create a sustainable, independent stream of income that might ultimately free the neighborhood from its dependency on outsiders.[8] As the trust grew, it could use the leverage that comes with wealth—a leverage that the rich enjoy as a matter of course—to restore the safety net that neoliberalism has shredded. It could, for instance, pressure would-be employers who wanted to contract with businesses and organizations in the trust to hire local firms, pay a living wage, and provide childcare, health care, and retirement benefits for their employees.[9] In addition, the trust could devote a small fraction of its resources to defray short-term costs to low-income residents and businesses, much like the old community chests that thrived in the first half of the twentieth century.[10]

The greatest challenge for our neighborhood trust would be acquiring the land to bring it under resident control. In many locations, particularly those on the cusp of gentrification, even vacant and abandoned land may be far too expensive for the residents of a distressed neighborhood to acquire, as landowners simply hold on to the unused land in the hope that they can sell it when values go up. In Dudley, the DNI acquired most of its land by a combination of eminent domain and direct transfer from the city. Wherever municipalities own vacant and abandoned property, they should transfer it to the ownership and control of a properly organized neighborhood trust. In places where this is not possible, however, the inability to purchase land makes it difficult for the trust to achieve its full potential.

To meet this challenge, I would add another tool that distinguishes a neighborhood trust from a community land trust. Turning neoliberalism into a tool that works *for* rather than against the low-income residents of a distressed neighborhood, I would increase the funding for neighborhood trusts by creating a combination of federal and state tax credits. State and federal governments could give neighborhood trusts a tax credit that they could sell on the open market, much in the way developers sell Low-Income Housing Tax Credits. Eligibility for the tax credit would be limited to neighborhood trusts that meet certain criteria, such as accounting transparency, demonstrated engagement with and responsiveness to area residents, and enforceable rules against self-dealing by trustees.

Investors who purchased the credits would receive a guaranteed stream of income for a term of years, ideally several decades or longer. The trust would use the proceeds of the sale to fund its operations and build its endowment, deploying the money to meet the overlapping needs of a distressed neighborhood without sacrificing long-term affordability. But unlike with the Low-Income Housing Tax Credit, neither investors nor state, federal, or local governments would have any control over the trust or its assets. If, for instance, neighborhood residents decided the best use of the money was to create single-room-occupancy housing for the lowest-income residents—or those who were homeless—then investors and the government could not object. As long as the assets were not used to discriminate against a protected class—by excluding people based, for instance, on race, ethnicity, religion, gender, or sexual orientation—the trust would have complete control over its funds. For the first time in U.S. history, the interests of capital would thus be aligned with the interests of the poor rather than the wealthy. Capital would grow, but people who have historically been marginalized and excluded would have control over the assets.

Though I believe tax credits are the ideal way to fund a neighborhood trust, they are not the only way. Cities have long used a host of tools to raise money for projects in low-income neighborhoods. Tax increment financing, for instance, allows a municipality to pool the property tax revenue generated within a designated district and use it to fund projects within the area. In addition, as community land trusts have gained popularity, supportive municipalities have begun to fund them in creative ways, including start-up grants, operating subsidies, and technical assistance.[11] Private institutions like Citibank, in partnership with nonprofits dedicated to affordable housing, have also contributed funds to create or expand existing CLTs.[12] All these tools should be used to fund neighborhood trusts. With any funding mechanism, however, the key condition, *which must not be sacrificed*, is to align the interests of capital with the interests of the poor; capital must be allowed to grow, since that ensures political viability in a neoliberal age, but the poor must own and control the assets.[13]

Nothing about a neighborhood trust would require neighborhood residents to break new organizational or legal ground. On the contrary, the trust would draw from established models, including community land trusts like DNI and nonprofit community development corporations like ONE Neighborhood Builders. Like DNI, the trust would acquire property and sell it only pursuant to a shared-equity arrangement. The rights reserved by the trust would be held in common by neighborhood residents and managed by the trustees. Like ONE Neighborhood, the trust would serve as a backbone organization for all funds that entered the neighborhood, distributing the resources and taking the lead in developing and maintaining relationships among the organizations that provided services within the neighborhood, some of which the trust itself could develop. By combining these two models, the neighborhood trust could develop a cradle-to-grave blanket of social services that tended to the overlapping and interconnected needs of low-income residents throughout their lives, while keeping the neighborhood affordable. And funding the blanket through tax credits uses a tool that is familiar to both lawmakers and investors.

In one respect, the neighborhood trust is the next stage in a long history of neighborhood development. It vests ownership and control with neighborhood residents rather than with outsiders, creates local expertise while building social capital and neighborhood wealth, and protects long-term affordability. By aligning capital with the poor rather than against them, the trust corrects the flaws of neoliberalism and allows a distressed neighborhood to raise itself from precarity without paving the way for displacement. In another respect, however, the neighborhood trust is the realization of a very old ambition. The current nonprofit sector came into existence more than a half century ago with the dream of giving low-income residents local ownership and control of neighborhood resources. But prior attempts have come to naught, and in today's neoliberal age, local ownership and control seem like a fantasy. The goal of the neighborhood trust is to make that fantasy a reality.

Like any innovation, the neighborhood trust will meet with objections. Some, for instance, will complain that my proposal is not radical enough. It takes neoliberalism as a given—as an unalterable background philosophy that writes the rules of the game—and merely tries to protect places like Olneyville from its toxic effects. Rather than call for neoliberalism to be ripped out by the roots, the trust deploys the tools of neoliberalism to fund its operations. This is a fair criticism, and I would like nothing better than for neoliberalism to be replaced by a robust national commitment to the welfare of the poor. I believe, for instance, that the federal government should get back in the business of building public housing. The national refusal to provide and care for the most vulnerable among us is obscene, especially since it is neoliberalism that has made inequality worse and swelled the ranks of the working poor. Yet in today's political climate, the obscenity will not end anytime soon. As the political scientist Sanford Schram observed, "As much as people have good reasons to wish away neoliberalism," it is "the new normal[,] and its pervasive influence must be taken into account in pushing for political change."[14] In a neoliberal age, neighborhoods like Olneyville must do what they can to transform themselves. When solutions are at hand, there is no point in waiting for a savior that may never come.

And even if neoliberalism were to disappear, I would still encourage Olneyville to develop a neighborhood trust. The Economic Opportunity Act of 1964, which launched the war on poverty, called for "maximum feasible participation" by the poor. This philosophy, however, was quickly abandoned, and the poor became passive recipients of federal largesse overseen by others.[15] Restoring robust and equitable government funding to neighborhoods like Olneyville is necessary but not sufficient; those funds must be owned and controlled by the residents of the neighborhood and should be held within a legally protected instrument that allows residents to resist the invasion of unrestrained capital.[16]

Other concerns are more substantial. Some will argue that shared-equity arrangements of the sort I have endorsed deprive low-income residents of what has historically been the surest path in this country to the accumulation of wealth: ownership of property. Because homeowners in a shared-equity agreement do not own the plot of earth beneath them, when they sell their homes, they miss out on the windfall that can result from dramatic increases in the value of land. Losing the chance for this individual wealth is no small thing.

This argument is made even more urgent when we recognize that access to this wealth has never been color-blind.[17] Throughout much of American history, Blacks and Latinos have systematically been denied access to mortgages and ownership in areas where property values were increasing, so that they were far less likely to reap the financial benefit of homeownership. White homeowners have been able to live in these areas, and they have relied on this increase to fund a more privileged lifestyle. A large percentage of the wealth gap between whites and Blacks is the result of this differential access to homeownership.[18] How can we justify a financial instrument like a neighborhood trust that perpetuates this inequity? Why shouldn't people of color have the same opportunity for homeownership as whites?

This argument has great moral and economic force, but it is ultimately not a good reason to reject a neighborhood trust. First of all, the existence of such a trust does not prevent a qualified purchaser from buying any home she can afford, including one on the open market. No one is *forced* to buy a home in the trust. Second, homes in the trust are available only to people who could not have afforded a home at market rates. Without shared equity, they are priced out of ownership entirely, at least in Olneyville. The trust allows them to enter and remain in a neighborhood that would otherwise quickly become unaffordable. In that way, the trust increases the number of low-income residents who can become homeowners in desirable places, even if they lack the full bundle of rights enjoyed by residents who can buy a home outright. In addition, the trust allows homeowners to build a credit history so that they can qualify for a conventional mortgage and later purchase a home on the open market.

Third, the argument against shared equity misses the point of the trust, which is to maintain the long-term affordability of the neighborhood as a whole. The trust protects the shared wealth of the neighborhood by stepping away from our obsessive focus on individual profit. The person who buys a home in the trust makes a commitment to sustaining a thriving, diverse, *and* affordable neighborhood. In exchange for that commitment, she relinquishes the right to cash in on gentrification, which would make her home unaffordable to the next generation of low-income buyers. The trust is an investment in the *neighborhood* at the expense of the individual. People who want to make that investment and be part of such a community will pay the economic price, and those who do not are free to buy elsewhere.[19]

Finally, the flexibility of the trust rules means that the terms of the shared-equity arrangement can be adapted to fit the neighborhood's changing needs. The neighborhood may decide, for instance, to alter the terms of the trust by allowing homeowners to purchase full ownership rights from the trust at a below-market rate or acquire them outright after a certain period of uninterrupted residency — say, fifteen or twenty years. All these options can be written into the trust agreement. An incentive for long-term residency would encourage the owner to stay in the neighborhood, maintain the house at a high level, and contribute to the neighborhood so as to increase the value of her investment. The residents in one neighborhood may want to write its trust rules to eliminate the opportunity for the owner to acquire the land rights; in another neighborhood, the residents may want to provide such an opportunity after twenty years; in another, after ten. If a neighborhood wants, it can even vote the trust out of existence and liquidate its assets. Like the neighborhood it protects, the trust can last forever, but it can also have a life span. As with everything else, the decision should be left to the residents.

Others will point out that developing a sizable endowment will take years, particularly if the federal government does not use a tool like tax credits to encourage capital to invest in the neighborhood trust. This is undoubtedly true, which is why I do not see the trust replacing the current approach right away. In the immediate term, money would continue to flow into the neighborhood from outside funders like philanthropies, foundations, community banks, and the government. But the funding should be owned and controlled by the neighborhood trust, rather than the nonprofits. The trust could use the resources to fund nonprofits and to operate the same sorts of initiatives and organizations that currently exist. In short, life with a neighborhood trust may at first look a great deal like life without it. Over time, however, the trust will grow in assets and importance, until eventually neighborhoods will no longer have to rely on outside funders to remain affordable to residents of limited means.

Still others will claim that a neighborhood trust merely replaces a diverse set of small, specialized nonprofits with a single large nonprofit. Yet a neighborhood trust does not merely create an extra layer of bureaucracy. It gives residents ownership and control over the resources that flow into their neighborhood—a power they do not presently have—and exercising that power is vital. Residents will have to decide democratically who is eligible to vote and serve on the trust. They may allow all neighborhood residents to vote, or they could limit voting to residents who have lived in the neighborhood for a particular length of time. They may create a board position for one or two nonresident allies, so that people like Frank Shea or Jennifer Hawkins could serve. Or they might restrict membership to low-income residents. But this complexity is a benefit, not a drawback, since these decisions are inherent in self-government. The capacity for, and attachment to, civic participation that a trust engenders would carry over to other aspects of community life. Residents who are engaged in the ownership and control of their neighborhood are more likely to be engaged in their city, state, and country. Creating and operating a neighborhood trust would increase self-governance.

Close attention to the inherent value of self-governance also helps us respond to a different objection. Some will say, quite rightly, that

a neighborhood does not speak with a single voice. Though Olneyville is predominately low income and Latino, great racial, ethnic, and
class diversity nonetheless exists in the neighborhood. It is ridiculous,
and more than a little insulting, to think that all Olneyville residents
have the same view on issues like policing, housing, charter schools,
and so on.[20] Given this diversity, it is entirely possible that a majority of low-income residents in Olneyville would not want a neighborhood trust. Indeed, they may prefer to retain the current model.
I fully accept this possibility. The point, however, is that the poor
should be able to make their own choices. What they choose does
not matter as long as the choice is theirs, expressed in free and fair
balloting. Anything else is at best paternalistic, and at worst colonial.

Another objection concerns the availability of land. Some cities
are now so densely built that practically no land exists that a trust
could acquire, even with funds from a tax credit. I grant that in some
locations, the tsunami of unrestrained capital has raised the cost of
land so much that there is no longer any point in talking about maintaining affordability for low-income residents. But those neighborhoods are not the ones we are trying to preserve. In those places,
capital has already thoroughly displaced the great majority of low-
income residents, and we can have little hope of bringing them back.
The neighborhood trust is meant for low-income neighborhoods that
can still be preserved, and in those places, land is still available. As the
Lincoln Institute of Land Policy reported in 2018, vacancies remain at
"epidemic levels" in our older industrial cities. In 2017, for instance,
Detroit had over 120,000 vacant lots. To put it mildly, where a neighborhood trust is most needed, a want of land is not the problem.[21]

This leads to the final objection: residents of distressed neighborhoods are not up to managing their own affairs. They lack the skills
needed to oversee a trust and its endowment, and they cannot do the
job of the many nonprofit organizations that currently provide vital

services to places like Olneyville. The poor, goes the argument, need better-educated, better-placed outsiders to do this work for them. This, in fact, is what I heard when I presented my research to members of the Olneyville Collaborative. "Great idea, but they're just not ready."

I disagree, strongly. To begin with, by treating the residents of a low-income neighborhood as part of the problem rather than part of the solution, this objection exposes the stewardship mentality of the current approach. Many people need training when they start a job, and if low-income residents of a distressed neighborhood lack skills, they can certainly acquire them. In addition, those who level this objection seem to suppose that an employee must be qualified to fill every position in an organization. That is not remotely true. Like anyone else, residents of a distressed neighborhood can hire advisers and outside experts and make informed judgments based on their input. They can also retain much of the staff that currently works in the nonprofits throughout the neighborhood. The key point is that these choices should be left to low-income residents rather than to outsiders.

More importantly, when I hear such objections, I cannot help but think of the long history of neighborhoods like Olneyville. Even if we assume the people who live there cannot manage their own affairs—an assumption I very much doubt—that deficit was caused in no small measure by two generations of neglect and mismanagement by people with greater resources and power. We cannot in good conscience create a system that impoverishes a neighborhood, and then point to neighborhood poverty as an excuse to keep the system in place. The history of distressed neighborhoods like Olneyville offers a powerful argument in *favor* of local control, not against it. In any case, the great success of community land trusts across the country, including Dudley Neighbors Incorporated, suggests that any supposed skill deficits within distressed neighborhoods are more imagined than real, and certainly not irreversible.

In the spring of 2019, I published an article in the *Stanford Social Innovation Review* that laid out my vision for a neighborhood trust.[22] The article struck a chord with many readers who, like me, were frustrated with the current approach to neighborhood well-being and were especially concerned that in so many low-income neighborhoods, the reward for years of hard work is gentrification and displacement. One such reader was Brian Murray, a principal at Shift Capital, in Philadelphia. Shift is a social impact real estate development firm, meaning its goal is not simply to accumulate wealth but to have a positive effect in distressed neighborhoods. I met with Brian and his partners as they took the idea of a neighborhood trust forward.

Shift is located in Kensington, a badly distressed neighborhood in North Philadelphia and an epicenter of the opioid crisis. A special agent for the Drug Enforcement Administration once called Kensington "the largest open-air heroin market on the East Coast."[23] It boasts the cheapest and purest heroin in the country and draws users from across the region. A local pastor called it "the Walmart of heroin." In late 2018, a writer for the *New York Times Magazine* captured the scene:

> The streetlights were broken or dim, and the alleyways were dark. Most of the blocks were lined with two-story rowhouses, abandoned factories and vacant lots. Kensington Avenue, the neighborhood's main drag, was a congested mess of Chinese takeouts, pawn shops, check-cashing joints and Irish pubs. Missing-person posters hung from storefront windows. The dealers were all out in the open, calling out brand names, even handing out free samples. Many people smoked

crack or meth or injected heroin. They stuck needles in their arms, necks and the skin between toes. They were limp and nodding off. Some people lay on the ground looking dead.[24]

Yet Kensington, like Olneyville, is also home to thousands of low-income residents who simply want a safe, affordable place to live and raise their families. As in Olneyville, a host of underfunded nonprofits serve area residents, but their efforts are hamstrung by the same conditions that burden nonprofits in Olneyville and other distressed neighborhoods. And like Olneyville, Kensington—for all its challenges—stands on the cusp of gentrification. The neighborhood's southern border marks the advancing edge of a gentrifying wave that presses relentlessly north. When I toured the area in early 2019 with the team at Shift, we visited the street that marked the dividing line between the cranes, construction, and upscale restaurants to the south and the vacant lots and boarded-up businesses to the north. Capital has Kensington in its sights.

The commercial heart of the neighborhood is Kensington Avenue, and that is where Shift has made its bet. It has joined forces with Impact Services, a well-established nonprofit that operates much like Olneyville's ONE Neighborhood Builders, to create the Kensington Corridor Trust, the first neighborhood trust in the country. The hope is that the trust will transform and stabilize the area by decommodifying property on and around Kensington Avenue. Shift Capital has already purchased 1.5 million square feet of commercial space, which it plans to place in the trust. This will remove the land from the private market, ensuring its long-term affordability and guaranteeing that the neighborhood captures any increases in the land's value. Ultimately the goal is to create a vibrant, sustainable mix of affordable residential and commercial rental units along the avenue. On the street level, the trust will rent to high-quality, culturally desirable businesses that provide affordable goods and services and create secure, high-paying jobs; above the businesses, the trust will create affordable apartments.

The Kensington Corridor Trust has begun to draw sympathetic media coverage.[25] The trust decommodifies land for the long-term benefit of neighborhood residents, and its managers hope to scale its operations beyond the limited reach of most community land trusts. Perhaps as importantly, they recognize the great urgency in a place like Kensington, which stands literally on the edge of the back-to-the-city movement in Philadelphia. Neighborhoods like Kensington simply cannot wait for Congress to create a tax credit that could fund a neighborhood trust. Shift has responded to this urgency by seeking to fund the trust through philanthropic investments. Yet that approach entails a great risk: How will the trust guarantee resident ownership and control? What happens if the interests of funders diverge from those of neighborhood residents? Allowing outsiders to control the assets in the trust destroys the very reason for having it, which is to vest ownership and control of neighborhood assets in low-income residents.

There are other risks. What happens if the trust successfully stabilizes the commercial corridor? Inevitably its success will raise nearby residential rents and property values. Will that price out low-income residents and contribute to gentrification and displacement? In addition, the trust defines the neighborhood broadly to include people who live outside Kensington but have a professional interest in its sustainability, including sympathetic investors and real estate professionals. Because of this broader perspective, low-income residents do not—at least at present—have a majority of seats on the trust's board. Should a rupture occur between outsiders and residents, the trust could turn into just another vehicle that thwarts the will of the poor.

To its credit, the Kensington Corridor Trust understands these risks perfectly well. I spoke with Adriana Abizadeh, the trust's executive director, who told me the goal is to transfer majority control to neighborhood residents, an ambition that is making its way into the documents governing the operation of the trust. She also plans to expand the number of residents on the board and shrink the number of outsiders. In addition, to minimize the risk that the trust will raise residential rents, members of the trust board are working in close collaboration with neighborhood nonprofits that are using conventional approaches like LIHTC to create affordable housing. The hope is that LIHTC housing will happen at the same time that the commercial corridor is stabilized.

Another, perhaps even more promising, model is taking shape in the Midwest. Trust Neighborhoods, a community development intermediary based in Kansas City, has created a model for neighborhood trusts that distressed neighborhoods can set up and run themselves. It matches existing neighborhood organizations with the model and helps find outside capital to create and fund each trust. The trust then acquires vacant land and distressed residential property, which it decommodifies by removing it from the private market and holding it within the trust for the benefit of low-income neighborhood residents. As neighborhoods implement the model, Trust Neighborhoods provides ongoing technical support, along with continuing access to resources and connections. Finally, the trust documents guarantee that its assets will be owned and controlled by neighborhood residents and their longtime allies rather than by outside investors.

Importantly, Trust Neighborhoods does not want to block gentrification. Instead it wants to co-opt it to benefit the poor. The Trust Neighborhoods model supports low-income residents by developing a mixed-income neighborhood. In fact, the organization calls its model a MINT: a mixed-income neighborhood trust. In a MINT, market-rate tenants will live in units owned by the trust, and the rent they pay will be used to subsidize low-income tenants renting at below-market rates living nearby—perhaps upstairs in a duplex or next door. The hope is that this mix of low-income and market-rate tenants, all renting from the trust, will allow the model to protect entire neighborhoods. By February 2021, neighborhoods in Tulsa and Kansas City had successfully adopted the model and created a MINT. Trust Neighborhoods expects another five neighborhoods will do so by the end of the year, and is in discussions with representatives from more than one hundred other low-income neighborhoods across the country.

I am cautiously optimistic about Kensington Corridor Trust and Trust Neighborhoods. Both are making great efforts to transform low-income neighborhoods while ensuring they do not become victims of their own success. Both models decommodify property to help guarantee that land in the neighborhood will be affordable forever. Both try to create local wealth that will eventually be owned and controlled by low-income residents rather than by outside capital. Both hope to provide an affirmative answer to the question that motivates this book: Can a distressed neighborhood transform itself without setting the stage for its own destruction? Whether they succeed remains to be seen.

At the end of the day, the current approach to neighborhood well-being will not protect places like Olneyville. The poor need to own the resources that flow into their neighborhood, and defenses to protect them from footloose capital. This means removing assets from individual hands and placing them in the shared hands of neighborhood residents, who hold those assets in trust for the benefit of the group and insulate them from private profit making and political manipulation. In an age when the failure of unregulated capitalism is becoming more evident every day, we cannot implement options like the neighborhood trust quickly enough.

11

And Then What?

I wrote this book to answer a simple question: Can a distressed neighborhood be transformed but not destroyed?

In Olneyville, a small band of nonprofits has paired with the police department and the local elementary school in a valiant attempt to prove the answer can be yes. They have restored a toxic infrastructure without yielding to the pressure to privatize public space. They have built and decommodified hundreds of homes in an effort to ensure their long-term affordability. They have constrained law enforcement so that it does no more than the low-income residents demand. They have created a public school that anchors the neighborhood and resists the centrifugal pull of choice and charters. They work collaboratively to minimize the competitive scramble for limited resources. And they have partnered with sympathetic allies at the highest levels of city and state government. To the extent they have succeeded, they have done so because at every step they have resisted the pressures of neoliberalism. They have refused to privatize or to privilege individual choice over communal well-being. They have resisted the siren call of the private sector. As best they can, they have limited the effect of neoliberal precarity. They have done everything right.

And still it is not enough. The improvements in Olneyville have not gone unnoticed. The neighborhood is becoming whiter and more expensive, inequality is growing, and rents are rising. The neighborhood has become fashionable and therefore vulnerable. The change has begun. Even when everything works as well as one can hope, the tools available in a neoliberal age cannot save neighborhoods like Olneyville, just as they have not saved hundreds of similar neighborhoods all across the country. The funding is too meager, the pressures of capital too great. Unless we try something new, distressed urban neighborhoods cannot be transformed without setting the stage for their destruction.

The solution is to attempt the one thing that has never been done in this country: empower the poor rather than the wealthy. By itself, a neighborhood trust will not solve all the problems that confront a place like Olneyville. It will not stop climate change, revive the manufacturing sector, or close the racial wealth gap. It will not end racism, classism, or xenophobia. It will not magically make people tolerant. Yet it will give the poor ownership and control of the wealth that flows into their neighborhoods. That will allow them to permanently capture increases in the value of land, invest and manage their capital, and deploy their assets as they see fit. It will restore some of the wealth and social capital that have been drained away by decades of disinvestment and neglect. It will align capital with the interests of the poor rather than the wealthy. The idea that the poor might achieve the same control over their destiny that the wealthy expect as their birthright is a radical notion in this country, but that is the unapologetic ambition of the neighborhood trust.

And then what? If the poor could live where their lives were not precarious—where they were not constantly hunting for cheaper rents and safer streets; where their children did not frolic in a toxic dump; where a working mother did not have to choose between her family's rent and her child's medicine; where families could set down roots in a stable, vibrant neighborhood; and where the threat of gentrification and displacement did not hang over their heads like the sword of Damocles—think of what we could accomplish. For the first time, all would have an equal chance at a long, healthy life. Children would receive an education that prepared them for the demands of responsible and productive citizenship in a technological age. Poverty would no longer be a pathway to entanglement with the criminal justice system. All the artificial distinctions between rich and poor would disappear. With their lives no longer precarious, people would finally be free to achieve their fullest potential.

And then what? What happens when the neighborhood trust is taken to scale and the working poor become an organized force, prepared and equipped to make demands on an economic and political system that has always exploited them? The poor have always vastly outnumbered the rich in this country. If the poor speak as one, they have the power to redistribute wealth, create a universal basic income, guarantee health care, and make housing a right. They can change U.S. policy on immigration, climate change, criminal justice, land use, and taxes. They can change the country. They don't need anyone to do it for them; they can do it themselves. And that is the unapologetic ambition of this book.

Epilogue

Olneyville's Walking School Bus stopped running in early 2020. Yuselly Mendoza graduated from college and took a part-time job with ONE Neighborhood as a community health worker. Another neighborhood leader and active supporter of the bus moved back to her native Costa Rica. When other parents couldn't afford to step in, the bus ground to a halt. Because nonprofits are chronically starved for resources, they inevitably rely on residents, who are expected to contribute their time for nothing or at most receive a modest stipend. Just as Delia Rodriguez-Masjoan warned, the model is unsustainable. Over the summer, Brent Kerman told me in an email that they expected to get the bus running again in the fall. But, he added, "Who knows what the fall will bring?"

I was completing this book in March 2020 when the greatest threat to neighborhood well-being in more than a century descended on the country. Within months, it was clear we would feel the pain of COVID-19 for years to come. And as always in this country, people of color and the poor have suffered disproportionately. In April 2020, the unemployment rate soared from 4.4 percent to 14.7 percent, the highest rate and sharpest one-month increase since the Great Depression. More than 20 million people lost their jobs in a single month. But as bad as these numbers are, they mask even greater pain among Blacks and Latinos. The Latino unemployment rate climbed to 18.9 percent. Among adult Latinas, it topped 20 percent. While the numbers were marginally better for Blacks, they were still worse than for whites.[1] These differences reveal the reality of employment segregation in this country. Blacks and Latinos are overrepresented in the low-wage trades that have been hit hardest by the economic shutdown: construction, hospitality, restaurants, retail, and cleaning and janitorial.[2]

For many of these low-wage workers, the financial precarity of neoliberalism is coming home to roost. According to the Federal Reserve, nearly 40 percent of Americans in 2019 did not have enough cash on hand to cover an unexpected expense of $400.[3] At the end of 2019, after the longest economic expansion in U.S. history, three in ten adults said that if they lost their job, they could not cover three months' expenses, even if they borrowed money.[4] And this was before the virus struck. By November 2020, food banks across the country were collapsing under the weight of unprecedented demand as hungry Americans lined up by the thousands for a bag of free food. At makeshift distribution centers that sprang up nationwide, the queue of cars stretched for miles.[5]

Just as Blacks and Latinos are overrepresented among those who have lost their source of income, they are overrepresented among the workers who must expose themselves to the risk of infection. These are the low-wage workers whose labor has been deemed essential or who cannot work remotely: delivery drivers, grocery and food service workers, health-care and hospital staff, and a range of other blue-collar employees.[6] Yet among all workers, they are least likely to have medical and sick leave benefits. The Bureau of Labor Statistics reported in 2019 that 31 percent of private-sector workers did not have access to medical benefits; the number soars to 57 percent for private-sector workers in service jobs. Nearly half the workers in the lowest wage quartile get no sick leave at all. And yet during the worst pandemic in more than a century, they are the people who are most likely to continue working outside their homes so that society can function, and thus most likely to become ill.[7]

Finally, the poor are more likely than the wealthy to live in neighborhoods like Olneyville, where they must contend with substandard housing, food insecurity, blight, and high rates of crime and police misconduct. Given these conditions, we should not be surprised that the poor also have higher rates of asthma, diabetes, hypertension, obesity, and heart disease, all of which increase their risk of dying from the virus should they become infected.[8]

These three factors—greater financial strain, less ability to shelter in place, and more preexisting risk factors—help explain why infection and mortality rates in Black and Latino neighborhoods are vastly higher than in white neighborhoods. In Iowa, for instance, Latinos represent only 6 percent of the population but account for more than 20 percent of COVID-19 cases. In Washington State, they make up 13 percent of the population but nearly a third of all cases.[9] Other places show similar disparities for Blacks. In Illinois, Blacks account for 15 percent of the population but 28 percent of all cases and 43 percent of all fatalities. In Michigan, only 14 percent of state residents are Black, but they account for a third of all cases and 40 percent of the deaths. In Louisiana, Blacks account for 32 percent of the population but 70 percent of COVID deaths.[10]

Rhode Island has not been spared either the virus or its disparate impact. In early May 2020, only three states had more cases per 100,000 residents (New York, New Jersey, and Massachusetts), and only seven states had more per capita deaths. Providence had by far the most cases and most deaths statewide, and only the badly distressed town of Central Falls had a higher rate of infection.[11] The Rhode Island Department of Health tracks cases carefully, but at least as of spring 2021, the department maps the spread of the disease by zip code rather than by neighborhood. One zip code stands out as the state's epicenter, with 10 percent of the statewide total and a per capita infection rate higher than New York City's. It is the zip code for Olneyville and two adjoining neighborhoods, Silver Lake and Manton.[12] And Olneyville has suffered one blow in particular from which it will not soon recover. In January 2021, Abelardo Hernandez succumbed to the virus. It is a monumental loss, both to his loving and tight-knit family and to the countless Olneyville residents on whose behalf he labored so long.

If COVID-19 teaches us anything, it is that we need a robust commitment to public health. Yet in a neoliberal age, that commitment has shriveled. The Centers for Disease Control is the federal agency tasked with protecting the country from public health threats. Over half of the CDC's budget goes to state and local agencies to support initiatives like the Walking School Bus. Adjusted for inflation, this portion of the CDC's budget has not changed much since 2008.[13] Over the same period, however, the need for health initiatives in neighborhoods like Olneyville has expanded significantly. Meanwhile the CDC's budget for public health emergencies like COVID-19 has been slashed by about a third since 2003.[14] In 2017, only 2.5 percent of the $3.36 trillion dollars spent on health care in the United States supported public health. In 2018, the nonpartisan Trust for America's Health calculated that public health in the United States was underfunded by $4.5 billion dollars. In 2019, the trust warned that "with chronic underfunding putting lives at risk, the stakes are rising."[15] COVID struck just months later.

The hemorrhaging of public health funding was most severe right after the Great Recession. Given the enormous impact of COVID on state and local coffers, funding will almost certainly be slashed again, and not just for public health. Across the country, states anticipate huge revenue shortfalls in 2020. Illinois expects its revenues will fall by more than $2 billion. South Carolina projects tax revenues alone to decline by over $500 million, or more than 15 percent. Personal income tax revenue in Pennsylvania came in nearly $1.5 billion below estimates, and the state calculates that $105 million was "permanently lost due to reduced economic activity."[16] Cities are in the same straits. Madison, Wisconsin, expects a $30 million revenue shortfall, Buffalo $36 million, San Francisco up to $1.5 billion. Nationwide, cities plan to slash services and furlough staff. In some places, the cuts have already begun.[17]

All of this means that the demands placed on nonprofits that support neighborhoods like Olneyville will increase exponentially, just as the funding they rely on shrinks. Many nonprofits will be lucky to survive. When COVID-19 was still in its earliest days, more than two-thirds of nonprofits had already seen a drop in contributions, and three in four had canceled or postponed fundraising.[18] When the dust finally settles, it is hard to imagine the nonprofit landscape will look remotely as it did at the start of 2020. Yet the damage done when nonprofits cut services or close their doors is felt most acutely not by the staff but by the low-income residents who rely so heavily on the services that these organizations provide.

COVID-19 did not create the inequities in American society, but it has magnified them and intensified their impact. As importantly, it has exposed the worst features of neoliberalism. Those whom neoliberalism made wealthy can protect themselves from the virus; they can work remotely, pay their utilities on time, have groceries delivered to their doorstep, and afford medical treatment should they become infected. Those whom neoliberalism left behind must bear the greatest risk with the least protection. For decades, neoliberals told us that the private sector would solve our problems and we should starve the government. Yet when disaster struck, the private sector could not save us but turned to the government with hands outstretched. Years of neoliberal austerity have left their mark, and states and cities are deep underwater. What remains of a safety net may soon be shredded. The nonprofit institutions we built to fill the gap created by the government's retreat will be overwhelmed. And in the end, it is the poor who will suffer. We can be sure that when that time comes, neoliberals will blame the poor for their own predicament.

The low-income residents of distressed neighborhoods cannot count on the world of neoliberal precarity to save them. They cannot count on capital or politicians. They cannot count on well-intentioned outsiders. They can only count on themselves. They must own the assets around them, take property out of the market and transfer it to the neighborhood, and seize control of their future, lest someone else seize it for them.

Notes

Prologue

1. "Neighborhood" is also preferable to larger units, like zip code or county. See, e.g., Adam Eckerd, "Cleaning Up without Clearing Out? A Spatial Assessment of Environmental Gentrification," *Urban Affairs Review* 47, no. 1 (January 2011): 31–59. ("Potential residents consider the reputation and quality of established neighborhoods, not tracts or zip codes.... Thus, the ideal level of analysis for tracking neighborhood change is the neighborhood itself.")

Chapter One. "Today Was a Good Day"

1. Bruce Katz and Jeremy Nowak, *The New Localism: How Cities Can Thrive in an Age of Populism* (Washington, D.C.: Brookings Institution, 2017). On the role of private capital in the current model, see, e.g., Thomas Miller, "Bridges to Dreams: The Story of the Low Income Investment Fund, Celebrating 25 Years of Impact, 1984–2009," Low Income Investment Fund, http://liifund.org/wp-content/uploads/2011/03/Bridges-to-Dreams-The-Story-of-LIIF-2009_LRes.pdf.

2. Incite!, *The Revolution Will Not Be Funded: Beyond the Nonprofit Industrial Complex* (Durham, N.C.: Duke University Press, 2017).

3. See, e.g., Thomas Sugrue, *The Origins of the Urban Crisis: Race and Inequality in Postwar Detroit* (Princeton, N.J.: Princeton University Press, 2005); Jessica Trounstine, *Segregation by Design: Local Politics and Inequality in American Cities* (Cambridge: Cambridge University Press, 2018).

4. See, e.g., David Harvey, *A Brief History of Neoliberalism* (New York: Oxford University Press, 2005).

5. See, e.g., Joshua Cohen, "Economics after Neoliberalism," *Boston Review,* July 29, 2019. ("What [Milton] Friedman calls 'liberalism' is the market fundamentalism that is now commonly called 'neoliberalism.'")

6. Philip Mattera and Kasia Tarczynska, "Uncle Sam's Favorite Corporations:

Identifying the Large Companies That Dominate Federal Subsidies," *Good Jobs First,* Mar. 2015, http://www.goodjobsfirst.org/sites/default/files/docs/pdf/UncleSamsFav oriteCorporations.pdf.

7. See, e.g., Stephen J. McGovern, "Philadelphia's Neighborhood Transformation Initiative: A Case Study of Mayoral Leadership, Bold Planning, and Conflict," *Housing Policy Debate* 17, no. 3 (2006): 529–570.

8. For a concise history of place-based community development strategies since the 1960s, see, e.g., Margery Austin Turner et al., "Tackling Persistent Poverty in Distressed Urban Neighborhoods: History, Principles, and Strategies for Philanthropic Investment," Urban Institute, June 2014, https://d2dmd73kpz2ss5.cloudfront.net /uploads/media_item/attachment/562e3e766e65696787110000/a66b798a-b3c2-47a6 -8d5c-378fb315410a.pdf; see also John Arena, *Driven from New Orleans: How Nonprofits Betray Public Housing and Promote Privatization* (Minneapolis: University of Minnesota Press, 2017). ("Part of the explanation [for the growth in nonprofits] is in the neoliberal transformation of the state.... As the state—or at least its social service arm—has receded, nonprofits have emerged to fill the void.")

9. David C. Hammack, "Nonprofit Organizations in American History: Research Opportunities and Sources," *American Behavioral Scientist* 45, no. 11 (2002): 1638–1674, 1662–1664, https://doi.org/10.1177/0002764202045011004.

10. Brice McKeever, "The Nonprofit Sector in Brief," Urban Institute, Dec. 13, 2018, https://nccs.urban.org/publication/nonprofit-sector-brief-2018#the-nonprofit -sector-in-brief-2018-public-charites-giving-and-volunteering.

11. Claire Cain Miller, "The Relentlessness of Modern Parenting," *New York Times,* Dec. 25, 2018, https://www.nytimes.com/2018/12/25/upshot/the-relentless ness-of-modern-parenting.html?action=click&module=Top%20Stories&pgtype =Homepage.

12. G. Wayne Miller, "$3.6M in Grants to Combat Health Disparities Announced by Rhode Island Foundation," *Providence Journal,* Oct. 19, 2018, https:// www.providencejournal.com/news/20181019/36m-in-grants-to-combat-health -disparities-announced-by-rhode-island-foundation—audio.

13. "LISC Rhode Island Awarded 455k to Seven Local Community Development Organizations in 2017," Local Initiatives Support Corporation Rhode Island, Feb. 27, 2018, http://rilisc.org/lisc-rhode-island-awarded-455k-to-seven-local-community-de velopment-organizations-in-2017-applications-for-2018-funding-open-friday.

14. Brett Theodos, Christina Plerhoples Stacy, and Helen Ho, "Taking Stock of the Community Development Block Grant," Urban Institute, Apr. 2017, https:// www.urban.org/sites/default/files/publication/89551/cdbg_brief_finalized_0.pdf.

15. Committee on Community-Level Programs for Youth et al., *Community Programs to Promote Youth Development,* ed. Jacquelynne Eccles and Jennifer Appleton Gootman (Washington, D.C.: National Academy Press, 2002).

16. United States Census Bureau, American Community Survey 2018 (5-Year Estimates), Census Tract 19, Providence County, Rhode Island, "Household Income Quintile, Upper Limits"; "Poverty Status in the Past 12 Months by Sex"; United States, "Poverty Status in the Past 12 Months by Sex." A church in Olneyville also provides food for two hours every Friday morning. See Rhode Island Food Bank, https://rifoodbank.org/find-food.

17. Though I suspect it will not be there much longer, in February 2021, the organization's website was still active. See English for Action, http://www.englishforaction .org.

18. Terms like "insider" and "outsider" are necessarily imprecise. I use "insider" as shorthand for the low-income minority residents who live in Olneyville. They are the people whose interests have been marginalized for decades, and whose place in the neighborhood is most threatened by gentrification and displacement. I recognize, however, that a person who lives outside Olneyville can be an ally to these residents, and some residents may not be allies. Despite this complexity, I maintain that the poor have a distinct set of interests that are threatened by neoliberalism, that those interests are not adequately protected by the current model of neighborhood well-being, and that the poor should be empowered to express their own views and control their own fate.

19. See, e.g., Richard Florida, "The Downsides of the Back-to-the-City Movement," *CityLab,* Sept. 29, 2016, https://www.citylab.com/equity/2016/09/downsides -of-the-back-to-the-city-movement/501476.

20. Indeed, it is hard to imagine a distressed urban neighborhood that does not experience *all* these problems to one degree or another, though the precise presentation will vary from place to place.

21. Fay Strongin, "You Don't Have a Problem Until You Do: Revitalization and Gentrification in Providence, Rhode Island" (MA thesis, City Planning, MIT, June 2017), https://dspace.mit.edu/bitstream/handle/1721.1/111259/1003291357-MIT .pdf?sequence=1.

22. Elizabeth Kneebone, "The Changing Geography of U.S. Poverty," testimony before the House Ways and Means Committee, Subcommittee on Human Resources, Feb. 15, 2017, https://www.brookings.edu/testimonies/the-changing-geog raphy-of-us-poverty/; Elizabeth Kneezone and Alan Berube, *Confronting Suburban Poverty in America* (Washington, D.C.: Brookings Institution Press, 2013); Aaron

Weiner, "Poverty Is Moving to the Suburbs. The War on Poverty Hasn't Followed," *Washington Post,* Apr. 5, 2018, https://www.washingtonpost.com/outlook/poverty-is -moving-to-the-suburbs-the-war-on-poverty-isnt-keeping-up/2018/04/05/cd4bc 770-3823-11e8-9c0a-85d477d9a226_story.html?utm_term=.955169e67ecd.

23. Elizabeth Kneebone and Natalie Holmes, "U.S Concentrated Poverty in the Wake of the Great Recession," Brookings Institution Report, Mar. 31, 2016, https:// www.brookings.edu/research/u-s-concentrated-poverty-in-the-wake-of-the-great -recession/; Ben Adler, "American Cities Have Bigger Things to Worry About Than Gentrification," *Grist,* Dec. 15, 2014, https://grist.org/cities/american-cities-have-big ger-things-to-worry-about-than-gentrification.

24. Jane Jacobs, *The Death and Life of Great American Cities* (New York: Random House, 1961); see also Louis Wirth, "Urbanism as a Way of Life," *American Journal of Sociology* 44, no. 1 (July 1938): 1–24; Steven A. Tuch, "Urbanism, Region, and Toler-ance Revisited: The Case of Racial Prejudice," *American Sociological Review* 52, no. 4 (1987): 504–510; J. Scott Carter, Lala Carr Steelman, Lynn M. Mulkey, and Casey Borch, "When the Rubber Meets the Road: Effects of Urban and Regional Resi-dence on Principle and Implementation Measures of Racial Tolerance," *Social Science Research* 34, no. 2 (2005): 408–425; Eric J. Oliver, *The Paradoxes of Integration: Race, Neighborhood, and Civic Life in Multiethnic America* (Chicago: University of Chicago Press, 2010); Christopher M. Huggins and Jeffrey S. Debies-Carl, "Tolerance in the City: The Multilevel Effects of Urban Environment on Permissive Attitudes," *Journal of Urban Affairs* 37, no. 3 (Aug. 2015): 255–269; Richard Florida, *The Rise of the Cre-ative Class—Revisited* (New York: Basic Books, 2012). Note that recent research sug-gests that urbanism promotes a tolerance of difference but not a tolerance of threat. See Huggins and Debies-Carl, "Tolerance in the City."

25. Lola Fadulu, "Hundreds of Thousands Are Losing Access to Food Stamps," *New York Times,* Dec. 4, 2019; "Final Rule: SNAP Requirements for Able-Bodied Adults without Dependents," U.S. Department of Agriculture, Food and Nutrition Service, Dec. 5, 2019, https://www.fns.usda.gov/snap/fr-120419.

Chapter Two. "Not a Particle of the Romantic"

1. Patrick T. Conley and Paul R. Campbell, *Providence: A Pictorial History* (Nor-folk, Va.: Donning, 1982), 100; Paul Davis, "R.I.'s Jewelry Industry in Search of a Permanent Home," *Providence Journal,* July 4, 2015.

2. Conley and Campbell, *Providence,* 100.

3. "Olneyville Community Contract," Olneyville Housing Corporation, Mar. 2010, 8.

4. "Olneyville Community Contract," 9.

5. William Morgan, "Olneyville's Mansion: Atlantic Mills," *GoLocalProv,* Aug. 26, 2016, https://www.golocalprov.com/business/Olneyvilles-Mansion-Atlantic-Mills.

6. Jenny R. Fields and Alyssa L. Wood, National Register of Historic Places Registration Form, Weybosset Mills Complex, Sept. 2007, http://www.preservation .ri.gov/pdfs_zips_downloads/national_pdfs/providence/prov_dike-street_weybosset -mills.pdf; Joseph D. Hall Jr., ed., *Biographical History of the Manufacturers and Businessmen of Rhode Island at the Opening of the 20th Century* (Providence, 1901), 298; *Olneyville Neighborhood Analysis* (City of Providence Department of Planning and Urban Development, 1979), 1, 6–13.

7. Conley and Campbell, *Providence,* 106.

8. *Olneyville Times,* Sept. 13, 1887. By the next year, Harris had dropped this line from his ads.

9. Norman Garrick, "Reimagining the Urban Freeway Holding Back Providence," *CityLab,* Nov. 30, 2016, https://www.citylab.com/transportation/2016/11/reim agining-the-urban-freeway-holding-back-providence/508373.

10. U.S. Congress, House of Representatives Committee on the Judiciary, "The Merger Movement in the Textile Industry," Staff Report, 84th Cong., 1st Sess., 1955, 5.

11. Alice Galenson, *The Migration of the Cotton Textile Industry from New England to the South: 1880–1930* (New York: Garland, 1985), 104–110.

12. See, e.g., H. E. Michl, *The Textile Industries: An Economic Analysis* (Washington, D.C.: Textile Foundation, 1938), 129–141; Michl, *The Textile Industries,* 138 ("The existence of too much productive capacity is the chief source of the industry's difficulties [and] more than any other single factor, is responsible for" chronic overproduction, low prices, irregular work, and low wages); New England Governors' Textile Committee, *The Merger Movement in the Textile Industry* (Cambridge, Mass., June 28, 1955), 4; Gary Gerstle, *Working Class Americanism: The Politics of Labor in a Textile City, 1914–1960* (New York: Cambridge University Press, 1989), 32–33; Steve Dunwell, *The Run of the Mill: A Pictorial Narrative of the Expansion, Dominion, Decline and Enduring Impact of the New England Textile Industry* (Boston: David R. Godine, 1978), 152.

13. Paul Buhle, "The Knights of Labor in Rhode Island," *Radical History Review* 17 (Spring 1978): 44–45, quoting "The Struggle for Existence," *People,* May 15, 1886.

14. Buhle, "The Knights of Labor," 42–46; David Vermette, "When an Influx of French-Canadian Immigrants Struck Fear into Americans," *Smithsonian,* August 21, 2019; "How the Italian Immigrants Came to New England," New England Historical Society, accessed February 18, 2020, https://www.newenglandhistoricalsociety .com/how-the-italian-immigrants-came-to-new-england; Jane Gerhard, "US Rub-

ber Company," Rhode Tour, accessed July 4, 2019, http://rhodetour.org/items/show /218; Taylor Polites, "West Side of Providence," Rhode Tour, accessed July 4, 2019, http://rhodetour.org/tours/show/22.

15. T. Jefferson Coolidge, *Chattanooga Tradesman*, Apr. 15, 1895, 83, quoted in Patrick J. Hearden, *Independence and Empire: The New South's Cotton Mill Campaign, 1865–1901* (DeKalb: Northern Illinois University Press, 1982), 92.

16. Beth Anne English, *A Common Thread: Labor, Politics, and Capital Mobility in the Textile Industry* (Athens: University of Georgia Press, 2006); Nancy Frances Kane, *Textiles in Transition: Technology, Wages, and Industry Relocation in the U.S. Textile Industry, 1880–1930* (Westport, Conn.: Greenwood Press, 1988); Galenson, *The Migration*.

17. Melvin Thomas Copeland, *The Cotton Manufacturing Industry of the United States* (Cambridge, Mass.: Harvard University Press, 1912), 27–28.

18. Galenson, *The Migration*, 39–46, 140–147.

19. Kane, *Textiles in Transition*, chap. 3.

20. Kimberly Amadeo, "Great Depression Timeline: 1929–1941," The Balance, https://www.thebalance.com/great-depression-timeline-1929-1941-4048064.

21. A. F. Hendricks, "Wage Rates and Weekly Earnings in the Cotton-Textile Industry, 1933–34," *Monthly Labor Review* 40, no. 3 (1935): 612–625.

22. Residential Security Map, Providence, R.I., 1935, National Archives, Record Group 195: Records of the Federal Home Loan Bank Board, 1933–1939, https://research.archives.gov/id/85713736.

23. James F. Findlay, "The Great Textile Strike of 1934: Illuminating Rhode Island History in the Thirties," *Rhode Island History* 42, no. 1 (Feb. 1983): 17–29; Rev. Edmund Joseph Brock, "A Study of Some Aspects of the Economic, Social, and Religious Life of a Selected Group of Young Catholic Workers" (M.S.W. thesis, Catholic University of America, 1938), ii–iii.

24. "Survey of Low-Rent Housing Needs as of September 1941, Providence, Rhode Island, WPA Project No. 40654, sponsored by the Housing Authority of the City of Providence, Rhode Island," Map 2, Inadequate Dwelling Units, Providence City Archives, Box HB 35, Old Pamphlets and Papers.

25. Michael Delaney, "Time Lapse: R.I. Shipyard Bustled with Workers in the 1940s," *Providence Journal*, Feb. 7, 2016.

26. Patrick J. Hearden, *Independence and Empire: The New South's Cotton Mill Campaign, 1865–1901* (DeKalb: Northern Illinois University Press, 1982); "New England's Textile Headache," *Challenge* 1, no. 1 (Oct. 1952): 38–47; John F. Kennedy, "New England and the South," *The Atlantic*, Jan. 1954.

27. "New England's Textile Headache."

28. United Transit Company, *A Study of Trade, Transit and Traffic: Providence, Rhode Island* (New Haven, Conn.: Wilbur Smith and Associates, 1958).

29. United Transit Company, *A Study of Trade,* 17–18, table 3; Conley and Campbell, *Providence,* 192.

30. Jane Gerhard, "US Rubber Company," Rhode Tour, http://rhodetour.org /items/show/218.

31. Conley and Campbell, *Providence,* 191.

32. The 1940 total is the highest recorded in a decennial census, though a Rhode Island census in 1925 showed a city population of nearly 268,000. See Conley and Campbell, *Providence,* 212.

33. Melih M. Ozbilgin and David Valletta, *Olneyville Neighborhood Analysis,* vol. 2 (City of Providence Department of Planning and Urban Development, 1979), 12, table II-1; U.S. Census, Decennial Digests.

34. Providence Redevelopment Agency, "Olneyville Square Revitalization Project: Proposed Redevelopment Plan, 1983," 2.

35. Wilbur Smith & Associates, "A Study of Trade, Transit and Traffic: Providence, Rhode Island," prepared for United Transit Company, Providence, New Haven, Conn., May 1958, 19–20, first quote from Robert W. Pratt, "Attitudes and Practices of Residents of Greater Providence concerning Downtown Providence: Report of a Survey for the Downtown Business Coordinating Council of the Greater Providence Chamber of Commerce," 1956.

36. Douglas S. Massey and Nancy A. Denton, *American Apartheid: Segregation and the Making of the Underclass* (Cambridge, Mass.: Harvard University Press, 1993), 42–57; Cassandra Netzke, "Rethinking Revitalization: Social Services in Segregation and Concentrations of Poverty," *Hamline Journal of Public Law and Policy* 23 (2001–2002): 145, 149–150.

37. "Warwick, Rhode Island, Population 2020," World Population Review, http:// worldpopulationreview.com/us-cities/warwick-ri-population/; "Cranston, Rhode Island Population 2020," World Population Review, http://worldpopulationreview .com/us-cities/cranston-ri-population.

38. Richard F. Weingroff, "Federal-Aid Highway Act of 1956: Creating the Interstate System," *Public Roads* 60, no. 1 (Summer 1996).

39. Conley and Campbell, *Providence,* 193, 198; Providence City Council Resolution No. 607, Approved Oct. 6, 1949; Ch. 2239, P.L., An Act Relating to Taking of Land and Easement by the City of Providence for Highway Purposes, passed by the Rhode Island General Assembly, January session, 1949. For the precise legal description of the land and property seized, see "Legal Notice of the Taking of Land

and Other Property for the City of Providence for the Olneyville Expressway," Oct. 18, 1949, Box 35, Providence City Archive.

40. Because complete records no longer exist, reconstructing exactly how many homes and businesses were destroyed is impossible. The city seized property within twelve different areas that formed a semicircle along the south and east side of Olneyville. Records in the Providence City Archives describe those twelve areas and indicate that 309 parcels of land were taken, some of which were vacant. John B. Carpenter, Inc., "Appraisal Areas A, B, C, D, J, K, L, M, Olneyville Expressway," in Box HB95, Providence City Archives, 1950; Assessment List, Olneyville Expressway, Areas G–H, Box HB95, Providence City Archives, n.d.

41. Carpenter, "Appraisal Areas."

42. Charles A. Maguire & Assocs., "The Olneyville Expressway, Providence, Rhode Island: What, Why, How, Where" (1949); see also Raymond A. Mohl, "The Interstates and the Cities: Highways, Housing, and the Freeways Revolt," Poverty and Race Research Action Council, 2002, 3, http://www.prrac.org/pdf/mohl.pdf (internal citation omitted). ("The victims of highway building tended to be overwhelmingly poor and Black. . . . It was quite obvious that neighborhoods and communities would be destroyed and people uprooted, but this was thought to be an acceptable cost of creating new transportation routes, facilitating economic development of the cities, and converting inner-city land to more acceptable or more productive uses. Highway builders and downtown redevelopers had a common interest in eliminating low-income housing and, as one redeveloper put it in 1959, freeing blighted areas 'for higher and better uses.'")

43. *Olneyville Neighborhood Analysis* (City of Providence Department of Planning and Urban Development, 1979), Providence City Archives, Box HB4, 20-A, chart 1, "Exterior Condition of Residential Structures."

44. *Olneyville Neighborhood Analysis.*

45. William Julius Wilson, *When Work Disappears: The World of the New Urban Poor* (New York: Knopf, 1996).

46. Quoted in Mike Stanton, *The Prince of Providence: The True Story of Buddy Cianci, America's Most Notorious Mayor, Some Wiseguys, and the Feds* (New York: Random House, 2003).

47. Douglas D. Perkins and D. Adam Long, "Neighborhood Sense of Community and Social Capital: A Multi-level Analysis," in *Psychological Sense of Community: Research, Applications, and Implications* (New York: Plenum, 2002), 291–318.

48. Wilson, *When Work Disappears,* 37–46.

49. Bryan Miller, "In Providence, Spring Rolls to Pizza," *New York Times,* Aug. 1, 1999, https://archive.nytimes.com/www.nytimes.com/library/travel/namerica/ct990 801.html; Carey Goldberg, "Bracing for Newest Round of Inquiry in Providence," *New York Times,* Mar. 19, 2000, https://www.nytimes.com/2000/03/19/us/bracing -for-newest-round-of-inquiry-in-providence.html; Ravi K. Perry, *21st Century Urban Race Politics: Representing Minorities as Universal Interests* (Bingley, U.K.: Emerald Group Publishing, 2013), chap. 1; Donna Heiderstat and Jennifer Chen, "World's Most Underrated Cities," *Travel and Leisure,* Mar. 1, 2011, https://www.travelandleisure .com/slideshows/worlds-most-underrated-cities?slide=105681#105681.

50. Mark T. Motte and Laurence A. Weil, "Of Railroads and Regime Shifts: Downtown Renewal in Providence, Rhode Island," *Cities* 17, no. 1 (2000): 7–18.

51. Motte and Weil, "Of Railroads and Regime Shifts." On the uneven effects of downtown revitalization, see, e.g., Marc V. Levine, "Downtown Redevelopment as an Urban Growth Strategy: A Critical Appraisal of the Baltimore Renaissance," *Journal of Urban Affairs* 9, no. 2 (1987): 103–123. Recent research in England has found that the high-tech jobs of the information economy actually *lower* the wages of less-skilled workers. Neil Lee and Stephen Clarke, "Do Low-Skilled Workers Gain from High-Tech Employment Growth? High-Technology Multipliers, Employment and Wages in Britain," *Research Policy* 48, no. 9 (Nov. 2019), https://www.sciencedirect .com/science/article/pii/S0048733319301234?via%3Dihub.

52. Mark McLaughlin, "The Disappearing Smokestack," *New England Business* 8, no. 5 (Sept. 15, 1986); see also William E. Collins, "Paolino's Agenda for Providence—Two Years Later," *Ocean State Business,* Feb. 13, 1989.

53. Motte and Weil, "Of Railroads and Regime Shifts," 7–18.

54. Laura Meade, "THE WAR ON DRUGS: Despite Lawmen's Efforts, R.I. Drug Problem Is 'Severe, Almost Epidemic,'" *Providence Journal,* May 15, 1988.

55. Collins, "Paolino's Agenda for Providence."

56. Laura Meade, "Police Attack Drugs in 2 More Areas," *Providence Journal,* June 2, 1989; Laura Meade, "4 Arrested in Raids; Drugs, Money Seized. Narcotics Squad Hits Two Olneyville Bars after Undercover Purchases of Cocaine," *Providence Journal,* Aug. 23, 1989; Laura Meade, "Boy, 16, Accused of Firing Gunshots. Resident: Dealers Terrorize Olneyville," *Providence Journal,* Oct. 24, 1990.

57. Rhode Island Governor's Justice Commission, "Highlights and Analysis from 25 Years of Collecting Serious Crime Data: A State, Cities and Towns Study, 1970–1994," Report No. 37, Sept. 1995; Rhode Island State Police, Uniform Crime Reporting Unit, "Crime in Rhode Island," Edition 43, 2000.

58. Bill Geller and Lisa Belsky, "A Study of Providence's Olneyville Neighborhood," in *Building Our Way Out of Crime: The Transformative Power of Police-Community Developer Partnerships* (Glenview, Ill.: Geller & Associates, 2012), 4–5, http://www.olneyville.org/Geller-Belsky-case-study.pdf.

59. Demetrios Caraley, "Washington Abandons the Cities," *Political Science Quarterly* 107, no. 1 (1992): 1–30.

60. Brett Theodos, Christina Plerhoples Stacy, and Helen Ho, "Taking Stock of the Community Development Block Grant," Urban Institute, Apr. 2017, https://www.urban.org/sites/default/files/publication/89551/cdbg_brief.pdf; Lissette Flores, "Trump Budget Eliminates Housing and Community Development Block Grants," Center on Budget and Policy Priorities, June 8, 2017, https://www.cbpp.org/blog/trump-budget-eliminates-housing-and-community-development-block-grants.

61. Barb Rosewicz, "'Lost Decade' Casts a Post-Recession Shadow on State Finances," Pew Charitable Trusts, June 4, 2019, https://www.pewtrusts.org/en/research-and-analysis/issue-briefs/2019/06/lost-decade-casts-a-post-recession-shadow-on-state-finances.

62. Rosewicz, "Lost Decade."

63. William Goldsmith, *Saving Our Cities: A Progressive Plan to Transform Urban America* (Ithaca, N.Y.: Cornell University Press, 2016), chap. 2.

64. For many years, it was impossible to calculate the value of tax breaks given to corporations for economic development. In 2015, however, the federal government required states to disclose the value of revenue lost to these abatements. Good Jobs First has compiled the total value of revenue lost, by state, for the years 2017–2019. See "State Tax Revenue Lost to Tax Abatement Programs 2017," Good Jobs First, https://taxbreaktracker.goodjobsfirst.org/prog.php?detail=top_state&order1=total_revenue_state_local&sort1=asc&fiscal_year=2017.

65. Steven Ruggles, Sarah Flood, Ronald Goeken, Josiah Grover, Erin Meyer, Jose Pacas, and Matthew Sobek, *IPUMS USA: Version 10.0* (census extracts for Providence, R.I., 1950, 1980, 2010) (Minneapolis: IPUMS, 2020), https://doi.org/10.18128/D010.V10.0.

66. Ruggles et al., *IPUMS USA.*

67. Marion Orr and Carrie Nordlund, "Political Transformation in Providence: The Election of Mayor Angel Tavares," in *21st Century Urban Race Politics: Representing Minorities as Universal Interests,* ed. Ravi K. Perry (Bingley, U.K.: Emerald Publishing Group, 2013), 1–12; Alexandra Filindra and Marion Orr, "Anxieties of an Ethnic Transition: The Election of the First Latino Mayor in Providence, Rhode Island," *Urban Affairs Review* 49, no. 1 (2013): 3–31, https://doi.org/10.1177/1078087412462449.

68. Melih M. Ozbilgin and David Valletta, *Olneyville Neighborhood Analysis*, vol. 2 (City of Providence Department of Planning and Urban Development, 1979), p. 15, table II-2, "Olneyville Population by White and Non-white, 1960 & 1970"; U.S. Census Decennial Digests.

69. Orr and Norlund, "Political Transformation in Providence."

70. "Biography of Councilwoman Sabina Matos," Providence City Council, https://council.providenceri.gov/sabina-matos-full-bio.

71. On the mobility of capital in the U.S. industrial context, see, e.g., Jefferson R. Cowie, *Capital Moves: RCA's Seventy-Year Quest for Cheap Labor* (Ithaca, N.Y.: Cornell University Press, 1999).

72. Susan S. Fainstain and Norman Fainstain, "New York City: The Manhattan Business District, 1945–1988," in *Unequal Partnerships: The Political Economy of Urban Redevelopment in Postwar America*, ed. Gregory D. Squires (New Brunswick, N.J.: Rutgers University Press, 1999), 61.

73. See David Koistinen, *Confronting Decline: The Political Economy of Deindustrialization in Twentieth-Century New England* (Gainesville: University Press of Florida, 2013), 5. ("Regional leaders were the key actors in creating the various responses to industrial decline in New England. Rank-and-file workers and other area residents were the ones most directly affected by the demise of long-established industries. But the efforts to counter deindustrialization were shaped by New England leaders and the institutions they guided.")

Chapter Three "We'll Pray for You"

1. See, e.g., Jon Shane, "The Problem of Abandoned Buildings and Lots," Center for Problem Oriented Policing, Arizona State University, July 2012, https://popcenter.asu.edu/content/abandoned-buildings-and-lots-0#endref10; William Spelman, "Abandoned Buildings: Magnets for Crime?" *Journal of Criminal Justice* 21, no. 5 (1993): 481–495. Note that Spelman's article focuses on abandoned residential buildings rather than factories.

2. Federal Emergency Management Agency, U.S. Fire Administration, Socioeconomic Factors and the Incidence of Fire, June 1997, 12–13.

3. "Five-Alarm Fire Sweeps Old Factory Complex," UPI, Dec. 18, 1989, https://www.upi.com/Archives/1989/12/18/Five-alarm-fire-sweeps-old-factory-complex/4890629960400.

4. Tom Middleton, "Raging Fire Sweeps through Old Factory Complex," UPI, Dec. 19, 1989, https://www.upi.com/Archives/1989/12/19/Raging-fire-sweeps-through-old-factory-complex/1226630046800.

5. "Five-Alarm Fire"; Middleton, "Raging Fire"; Michael F. Moise, obituary, *Hartford Courant,* Mar. 4, 1997, http://articles.courant.com/1997-03-04/news/9703 040310_1_fire-chief-lewis-marshall-fire-prevention-division-major-fire.

6. "Torching Tips Probed in Olneyville Mill Fire. Sprinklers Tampered With, Police Are Told," *Providence Journal,* Jan. 5, 1990; "Arson Investigators Comb Rubble of Riverside Mills Fire Classified as Suspicious Because of Its Speed," *Providence Journal,* Dec. 28, 1990.

7. "Woonasquatucket River Fecal Coliform Bacteria and Dissolved Metals Total Maximum Daily Loads," Rhode Island Department of Environmental Management, 2007, 18, http://www.dem.ri.gov/programs/benviron/water/quality/rest/pdfs/woofinal.pdf.

8. Bill Geller and Lisa Belsky, "A Study of Providence's Olneyville Neighborhood," in *Building Our Way Out of Crime: The Transformative Power of Police-Community Developer Partnerships* (Glenview, Ill.: Geller & Associates, 2012), 4–5, http://www.olneyville.org/Geller-Belsky-case-study.pdf.

9. Dean Isabella, "What Was Done, Was Done Right," LISC Rhode Island, Mar. 22, 2017, http://www.lisc.org/our-stories/story/what-was-done-was-done-right; Thomas E. Deller, "A Tale of Two Cities: Economic Development and Housing Change in Hartford, Connecticut, and Providence, Rhode Island," seventh annual lecture of the Dr. Marcia Marker Feld Lecture Series for Social Justice and the City, 2014.

10. Gary Rivlin, "The Night Chicago Burned," *Chicago Reader,* Aug. 25, 1988, https://www.chicagoreader.com/chicago/the-night-chicago-burned/Content?oid=872662.

11. Simon E. Balto, "MLK's Forgotten Plan to End Gun Violence in Chicago," *History News Network,* July 8, 2013; James Alan MacPherson, "Chicago's Blackstone Rangers," *The Atlantic,* May 1969.

12. Katherine Gregg, "Election Analysis: Loyalty, Discontent Helped Cianci," *Providence Journal,* Nov. 7, 1990.

13. Brown University, Office of the University Curator, Biography of Frederick Lippitt, n.d., http://library.brown.edu/cds/portraits/display.php?idno=314.

14. Mike Stanton, *The Prince of Providence: The True Story of Buddy Cianci, America's Most Notorious Mayor, Some Wiseguys, and the Feds* (New York: Random House, 2003).

15. Stanton, *The Prince of Providence,* 154.

16. Stanton, *The Prince of Providence,* 203.

17. Gregg, "Loyalty, Discontent Helped Cianci."

18. Gregg, "Loyalty, Discontent Helped Cianci"; Stanton, *The Prince of Providence*, 207.

19. Emily Gold Boutilier, "A Life of Service," *Brown Alumni Magazine*, Apr. 27, 2007. https://www.brownalumnimagazine.com/articles/2007-04-27/a-life-of-service.

20. In 2017, long after Sherman had left the organization, the plan's former finance director pleaded guilty to defrauding the group of three-quarters of a million dollars and was sentenced to just under three years in federal prison. The plan has since closed its doors. Department of Justice, U.S. Attorney's Office, District of Rhode Island, "Providence Plan Finance Director Sentenced for Embezzling $742,190," Aug. 11, 2017, https://www.justice.gov/usao-ri/pr/providence-plan-finance-director -sentenced-embezzling-742190.

21. Stanton, *The Prince of Providence*, 413.

22. Press release, "Governor Lincoln D. Chafee Announces Ten Appointments to Board of Governors for Higher Education," Office of the Governor, Feb. 16, 2011, http://www.ri.gov/press/view/13216.

23. On the relationship between crime and urban parks, and especially the importance of designing for safety, see, e.g., Lincoln Larson and S. Scott Ogletree, "How to Design a Park That Deters Crime," *CityLab* (June 26, 2019), https://www .citylab.com/life/2019/06/parks-crime-safety-green-space/592648 ("Parks that are designed for safety, heavily programmed on an ongoing basis, and well maintained tend to attract residents whose presence serves as a crime deterrent"); "What Role Can Design Play in Creating Safer Parks," Project for Public Spaces, Dec. 31, 2008, https://www.pps.org/article/what-role-can-design-play-in-creating-safer-parks.

24. See, e.g., Jim Hilborn, *Dealing with Crime and Disorder in Urban Parks* (United States Department of Justice, Office of Community Oriented Policing Services, 2009).

25. Merino Park, Woonasquatucket River Watershed Council, n.d., https:// wrwc.org/wp/what-we-do/greenway/parks/merino-park.

26. Academics who study the effectiveness of civic organizations distinguish between mobilizers and organizers. Mobilizers try to engage as many people as possible to support various initiatives, while organizers aim for numbers but in addition try to build relationships and develop new leaders and activists within the community. Organizers, in short, aim for depth and breadth, while mobilizers typically set their sights on breadth alone. See, e.g., Hahrie Han, *How Organizations Develop Activists: Civic Associations and Leadership in the 21st Century* (New York: Oxford University Press, 2014). By this metric, Lisa is a mobilizer.

27. William J. Clinton, "Address before a Joint Session of the Congress on the State of the Union," Feb. 4, 1997, online by Gerhard Peters and John T. Woolley, American Presidency Project, https://www.presidency.ucsb.edu/node/223396/; "Brownfields Showcase Community," Environmental Protection Agency, 1998.

28. "Federal Support of Community Efforts along American Heritage Rivers," *Federal Register* 62, no. 178 (Sept. 15, 1997), https://www.gpo.gov/fdsys/pkg/FR-1997 -09-15/pdf/97-24591.pdf.

29. "Past Honorees and Keynote Speakers," Environmental Law Institute, Oct. 28, 2015, https://www.eli.org/award-dinner/history.

30. Adam Clymer, "John Chafee, Republican Senator and a Leading Voice of Bipartisanship, Dies at 77," *New York Times,* Oct. 26, 1999, http://www.nytimes .com/1999/10/26/us/john-chafee-republican-senator-and-a-leading-voice-of-biparti sanship-dies-at-77.html.

31. G. Wayne Miller, "A Humble Path to Power," *Providence Journal,* Apr. 27, 2008; Jack Reed Senate website, https://www.reed.senate.gov/about/jacks-story.

32. Jane Sherman, "American Heritage Rivers Cover Sheet for Nomination Package," Dec. 10, 1997, https://semspub.epa.gov/work/01/490404.pdf; press release, "President Clinton: Celebrating America's Rivers," July 30, 1998, https://clinton whitehouse2.archives.gov/CEQ/Rivers/; "Brownfields Case Study: Woonasquatucket River Greenway Project," Rhode Island Department of Environmental Management, Apr. 3, 2003, http://www.dem.ri.gov/brownfields/successes/woonasquatucket.htm. Technically, Providence was designated a Brownfields Showcase Community, and two sites along the Woonasquatucket were chosen for remediation: Riverside Mills in Olneyville, and Lincoln Lace and Braid in a nearby neighborhood. See "Brownfields Showcase Community: State of Rhode Island / Providence, Quick Reference Fact Sheet," Environmental Protection Agency, Oct. 2000, https://nepis.epa.gov/Exe /tiff2png.cgi/P100BH3L.PNG?-r+75+-g+7+D%3A%5CZYFILES%5CINDEX%20 DATA%5C95THRU99%5CTIFF%5C00002539%5CP100BH3L.TIF.

33. "Brownfields Case Study."

34. News release, RI-Department of Environmental Management, Oct. 23, 1999. Thomas Deller, from the city planning department, told me he later secured another $2 million from HUD to help pay for the remediation. I have not been able to find any independent confirmation of that award, and other participants in the remediation process did not mention it.

35. EA Engineering, Science, and Technology, "Remedial Action Closure Report, Riverside Mills Restoration Project, 50 Aleppo Street, Providence, Rhode

Island, 02905," Nov. 2014, 6. I am grateful to Robert McMahon for unearthing this report from the Parks Department files.

36. EA Engineering, Science, and Technology, "Remedial Action Closure Report," 7.

37. The park officially opened in 2008. See "Riverside Park," Woonasquatucket River Watershed Council, https://wrwc.org/wp/what-we-do/greenway/parks/river side-park.

38. See, e.g., Peter Harnik and Ben Welle, "Measuring the Economic Value of a City Park System," Trust for Public Land, 2009, http://cloud.tpl.org/pubs/ccpe -econvalueparks-rpt.pdf; Beach Coccevilla, "A Smart Investment for America's Health," City Parks Alliance, 2015, https://cityparksalliance.org/wp-content/up loads/2019/06/CPA_LWCF_HEALTH.pdf?pdf=report; Myron Floyd et al., "Cost Analysis for Improving Park Facilities to Promote Park-Based Physical Activity," North Carolina Cooperative Extension, 2015 ("A growing number of studies suggest that parks contribute significantly to physical activity among adults and children"), https://content.ces.ncsu.edu/cost-analysis-for-improving-park-facilities-to-promote -park-based-physical-activity.

39. Much of this research is collected at the websites of the National Recreation and Park Association, http://www.nrpa.org/; the City Parks Alliance, http://www .cityparksalliance.org/index.php; and the Trust for Public Land, https://www.tpl.org. Scholars have developed other tools to evaluate the impact of remediating brown-fields generally. See, e.g., Laurel Berman, Christopher A. De Sousa, Terri Linder, and David Misky, "From Blighted Brownfields to Healthy and Sustainable Communities: Tracking Performance and Measuring Outcomes," in *Reclaiming Brownfields: A Comparative Analysis of Adaptive Reuse of Contaminated Properties,* ed. Richard C. Hula, Laura A. Reese, and Cynthia Jackson-Elmoore (New York: Routledge, 2016).

40. Aurecchia told me that the creation of Riverside Park was not her earliest collaboration with the Providence Police Department. While Jane Sherman was attempting to raise money and generate support for the creation of Riverside Park, Aurecchia joined with the Providence police and Olneyville residents in a campaign to rescue Donigian Park, or Needle Park, as it was known. Aurecchia organized residents to conduct massive cleanup operations in the park, and the police prevailed on residents to allow the police to film drug activity from residents' windows so that the police could arrest dealers. While this effort certainly laid a foundation for the subsequent collaboration at Riverside, the latter was the more significant event in the transformation of Olneyville. Riverside is much larger than Donigian and involved

the creation of a new park from what was once a toxic waste dump. In addition, the success at Riverside involved both the creation of the park and the construction of affordable housing along Aleppo Street, matters I take up in chapter 4. Finally, the experience at Riverside prompted Frank Shea, the executive director at Olneyville Housing Corporation, to revitalize the Olneyville Collaborative to facilitate future partnerships among neighborhood groups. In total, therefore, Riverside had a more dramatic effect on the landscape and engaged more members of the nonprofit community.

41. See, e.g., Peter Harnik and Ryan Donahue, "Turning Brownfields into Parks: On Once-Polluted Properties, Bold New Public Spaces Deliver the Green," Trust for Public Land, 2012, http://cloud.tpl.org/pubs/ccpe-brownfield-article-2012.pdf.

42. News release, "EPA Announces the Selection of 155 Grants for Communities," Environmental Protection Agency, May 20, 2020, https://www.epa.gov/newsreleases /epa-announces-selection-155-grants-communities-receive-over-65-million-total -grant.

43. "Brownfields Program Accomplishments and Benefits," Environmental Protection Agency, n.d., https://www.epa.gov/brownfields/brownfields-program-accom plishments-and-benefits. The EPA's estimate is likely far too low. See Gwyneth Shaw, Beverly Ford, and Evelyn Larrubia, "EPA's 'Brownfields' Program Coming Up Short," Center for Public Integrity, Sept. 25, 2012, https://publicintegrity.org/environment /epas-brownfields-program-coming-up-short.

44. Shaw, Ford, and Larrubia, "EPA's 'Brownfields' Program."

45. Shaw, Ford, and Larrubia, "EPA's 'Brownfields' Program."

46. "Anatomy of Brownfields Redevelopment," Office of Brownfields and Land Revitalization, Environmental Protection Agency, June 2019, https://www.epa.gov /sites/production/files/2015-09/documents/anat_bf_redev_101106.pdf.

47. "Anatomy of Brownfields Redevelopment," 1.

48. "Anatomy of Brownfields Redevelopment," 3. ("Public development" that remediates sites for public use, such as a park or a school, "most often occurs when there is little private-sector interest in a property because ... market factors ... limit potential development options.")

49. Solitare and Greenberg found that the cities receiving brownfield remediation funding from the EPA in the mid-1990s tended to be more distressed compared to the national average. Laura Solitare and Michael Greenberg, "Is the U.S. Environmental Protection Agency Brownfields Assessment Pilot Program Environmentally Just?" *Environmental Health Perspectives* 110, no. 2 (2002): 249–257. I would argue,

however, that looking at entire cities uses too broad a lens; the question is whether the funding goes to distressed neighborhoods within those cities. As the authors correctly noted, their study does not address this question. In any case, the authors also noted that funding began to go to considerably less distressed cities in the latter years of the pilot (250).

50. Shaw, Ford, and Larrubia, "EPA's 'Brownfields' Program."

51. "Anatomy of Brownfields Redevelopment," 1.

52. "Anatomy of Brownfields Redevelopment," 7.

53. "Anatomy of Brownfields Redevelopment," 6, 8–9.

54. Kate Golden, "Wasted Places: Slow, Underfunded EPA Program Falls Short in Toxic Site Cleanups," Wisconsin Center for Investigative Journalism, Sept. 23, 2012, https://www.wisconsinwatch.org/2012/09/arsenic-and-old-places-brownfield -cleanups-beset-by-flawed-program.

55. Jonathan D. Essoka, "The Gentrifying Effects of Brownfields Redevelopment," *Western Journal of Black Studies* 34, no. 3 (2010): 299–315; see also National Environmental Advisory Council, "Unintended Impacts of Redevelopment and Revitalization Efforts in Five Environmental Justice Communities," Aug. 2006, 2, https://www.epa.gov/sites/production/files/2015-02/documents/redev-revital-reco mm-9-27-06.pdf. ("From the perspective of gentrified and otherwise displaced residents and small businesses it appears that the revitalization of their cities is being built on the back of the very citizens who suffered, in-place, through the times of abandonment and disinvestment. While these citizens are anxious to see their neighborhoods revitalized, they want to be able to continue living in their neighborhoods and participate in that revitalization.")

56. See, e.g., Kenneth A. Gould and Tammy L. Lewis, *Green Gentrification: Urban Sustainability and the Struggle for Environmental Justice* (New York: Routledge, 2017); Melissa Checker, "Wiped Out by the 'Greenwave': Environmental Gentrification and the Paradoxical Politics of Urban Sustainability," *City and Society* 23 (2011): 210–229; Dan Immergluck, "Sustainable for Whom? Large-Scale Sustainable Urban Development Projects and Environmental Gentrification,'" *Shelterforce: The Voice of Community Development*, Sept. 1, 2017, https://shelterforce.org/2017/09/01/sustain able-large-scale-sustainable-urban-development-projects-environmental-gentrifica tion/; Juliana A. Maantay and Andrew R. Maroko, "Brownfields to Greenfields: Environmental Justice versus Environmental Gentrification," *International Journal of Environmental Research and Public Health* 15, no. 10 (2018): 2233.

Chapter Four. "We'll Be Happy to Indulge Your Fantasies"

1. Lauren Prescott, "People before Highways: The Southwest Expressway," *South End Historical Society Newsletter* 44, no. 2 (Fall 2016), http://www.southendhis toricalsociety.org/wp-content/uploads/2012/08/SEHS2016FALL_FINAL_web.pdf; "People before Highways," Jamaica Plain Historical Society, n.d., http://www.jphs .org/transportation/people-before-highways.html; "Southwest Corridor Park History," Southwest Corridor Park Conservancy, n.d., http://www.swcpc.org/history .asp; Catherine Foster, "Subway with a Park on Top. Boston's Southwest Corridor Has Train Tracks Below, Green Areas and Recreation Above," *Christian Science Monitor,* Feb. 24, 1989, https://www.csmonitor.com/1989/0224/dorange.html.

2. Prescott, "People before Highways"; "Southwest Corridor Park History"; Foster, "Subway with a Park."

3. "U.S. Census Bureau QuickFacts: Boston City, Massachusetts," Census Bureau QuickFacts, https://www.census.gov/quickfacts/bostoncitymassachusetts.

4. "Creating Safe Park Environments to Enhance Community Wellness," National Recreation and Park Association, https://www.nrpa.org/contentassets/f76842 8a39aa4035ae55b2aaff372617/park-safety.pdf.

5. Nabihah Maqbool, Janet Viveiros, and Mindy Ault, "The Impacts of Affordable Housing on Health: A Research Summary," Center for Housing Policy, Apr. 2015, https://www.rupco.org/wp-content/uploads/pdfs/The-Impacts-of-Affordable -Housing-on-Health-CenterforHousingPolicy-Maqbool.etal.pdf.

6. Maya Brennan, Patrick Reed, and Lisa Sturtevant, "The Impacts of Affordable Housing on Education: A Research Summary," Center for Housing Policy, Nov. 2014; Maya Brennan, "The Impacts of Affordable Housing on Education: A Research Summary," Center for Housing Policy, May 2011, https://www.enterprisecommunity .org/download?fid=7946&nid=3582.

7. Ayoung Woo and Kenneth Joh, "Beyond Anecdotal Evidence: Do Subsidized Housing Developments Increase Neighborhood Crime?" *Applied Geography* 64 (Oct. 2015): 87–96; Matthew Freedman and Emily G. Owens, "Low-Income Housing Development and Crime," *Journal of Urban Economics* 70, nos. 2–3 (Sept.–Nov. 2011): 115–131.

8. Matthew Desmond, "How Home Ownership Became the Engine of American Inequality," *New York Times Magazine,* May 9, 2017, https://www.nytimes.com /2017/05/09/magazine/how-homeownership-became-the-engine-of-american-in equality.html; see also Matthew Desmond, "No Place like Home: America's Eviction Epidemic," *The Guardian,* Feb. 12, 2017, https://www.theguardian.com/society/2017 /feb/12/americas-eviction-epidemic-matthew-desmond-housing-crisis.

9. Spencer Agnew, "The Impact of Affordable Housing on Communities and Households," Minnesota Housing Finance Agency, n.d., http://www.mhponline.org/files/AffordableHousingImpact-CommunitiesandHouseholds.pdf; see also Rebecca Cohen, "The Positive Impacts of Affordable Housing on Health: A Research Summary," Center for Housing Policy, 2007, https://www.enterprisecommunity.org/download?fid=8265&nid=4141.

10. Joint Center for Housing Studies, "The State of the Nation's Housing: 2019," https://www.jchs.harvard.edu/sites/default/files/Harvard_JCHS_State_of_the_Nations_Housing_2019.pdf; National Low Income Housing Coalition, "The GAP: A Shortage of Affordable Homes," Mar. 2020, https://reports.nlihc.org/sites/default/files/gap/Gap-Report_2020.pdf.

11. See Michael E. Stone, "Housing Affordability: One-Third of a Nation Shelter Poor," in *A Right to Housing: Foundation for a New Social Agenda,* ed. Rachel Bratt, Michael E. Stone, and Chester Hartman (Philadelphia: Temple University Press, 2006); U.S. Census Bureau, American Community Surveys, 2018 (5-Year Estimates), "Poverty Status in Families by Family Type by Presence of Children under 18 Years"; "Children Living with Single Parents."

12. See, e.g., "6 Reasons That Pay Has Lagged behind U.S. Job Growth," *New York Times,* Feb. 1, 2018.

13. See, e.g., Sanford F. Schram, *The Return of Ordinary Capitalism: Neoliberalism, Precarity, Occupy* (New York: Oxford University Press, 2015), chap. 1; Jay Shambaugh, Ryan Nunn, Patrick Liu, and Greg Nantz, "Thirteen Facts about Wage Growth," the Hamilton Project, 2017.

14. "Worker Rights Preemption in the U.S.," Economic Policy Institute, Nov. 2018, https://www.epi.org/preemption-map.

15. United States Census Bureau, American Community Survey 5-year estimates, 2011–2015; United States Census Bureau, "Housing Cost of Renters: 2000," May 2003, https://www.census.gov/prod/2003pubs/c2kbr-21.pdf.

16. Fay Strongin, "You Don't Have a Problem Until You Do: Revitalization and Gentrification in Providence, Rhode Island" (MA thesis, City Planning, MIT, June 2017), https://dspace.mit.edu/bitstream/handle/1721.1/111259/1003291357-MIT.pdf?sequence=1.

17. Martha Ross and Nicole Bateman, "Meet the Low-Wage Workforce," Metropolitan Policy Program at Brookings, Nov. 2019, https://www.brookings.edu/wp-content/uploads/2019/11/201911_Brookings-Metro_low-wage-workforce_Ross-Bateman.pdf#page=9.

18. "America's Middle-Class Meltdown: Fifth of US Adults Live in or near to

Poverty," *Financial Times,* Dec. 11, 2015, https://www.ft.com/content/c3de7f66-9f96-11e5-beba-5e33e2b79e46.

19. United States Census Bureau, American Community Surveys (5-Year Estimates), Census Tract 19, Providence County, Rhode Island, "Income in the Past 12 Months in 2015 and 2018," https://data.census.gov/cedsci/table?q=income&hidePreview=true&tid=ACSST5Y2018.S1901&vintage=2018&g=1400000US44007001900.

20. See, e.g., Katherine S. Newman, *No Shame in My Game: The Working Poor in the Inner City* (New York: Vintage, 2000); Arne L. Kalleberg, "Precarious Work, Insecure Workers: Employment Relations in Transition," *American Sociological Review* 74, no. 1 (Jan. 2009): 1–22. For a review of the academic literature on employment precarity and its link to neoliberalism, see Arne L. Kalleberg and Steven P. Vallas, "Probing Precarious Work: Theory, Research, and Politics," in *Precarious Work, Research in the Sociology of Work,* vol. 31, ed. Arne L. Kalleberg and Steven P. Vallas (Bingley, U.K.: Emerald Publishing, 2018), 1–30, http://arnekalleberg.web.unc.edu/files/2018/01/Precarious-Work-CH-1.pdf. ("Evidence from a variety of diverse objective indicators generally supports the view that there have been reductions in social and statutory protections associated with employment relations and an increase in overall job insecurity in the United States.... Put differently, neoliberalism has pressured both nation-states and capital to uproot those organizational arrangements that had shielded workers from the vicissitudes of the external labor market.")

21. Jason DeParle, "Slamming the Door," *New York Times,* Oct. 20, 1996.

22. "First-Ever Evictions Database Shows: 'We're in the Middle of a Housing Crisis,'" National Public Radio, Apr. 12, 2018, https://www.npr.org/templates/transcript/transcript.php?storyId=601783346.

23. Charles Bagli, "In Program to Spur Affordable Housing, $100 Million Penthouse Gets 95% Tax Cut," *New York Times,* Feb. 1, 2015; Janet Babin, "Billionaire's Boondoggle or Essential Tax Credit," WNYC, Jan. 29, 2015, https://www.wnyc.org/story/billionaires-boondoggle-or-essential-tax-credit-more-debate-over-421-/; Ross Barkan, "Progressives March against Tax Break for Luxury Developments," *Observer,* Feb. 18, 2015, http://observer.com/2015/02/progressives-march-against-tax-break-for-luxury-developments/; Rowley Amato, "Which Luxury Apartments Are Getting 421-a Tax Abatements," *Curbed,* Mar. 24, 2015, https://ny.curbed.com/2015/3/14/9980782/which-luxury-buildings-are-getting-421-a-tax-abatements.

24. Alex F. Schwartz, *Housing Policy in the United States* (New York: Routledge, 2015), 47.

25. Schwartz, *Housing Policy in the United States,* 55; "Policy Basics: Public Hous-

ing," Center on Budget and Policy Priorities, Nov. 15, 2017, https://www.cbpp.org /research/policy-basics-public-housing.

26. Alana Semuels, "America's Shame: How U.S. Housing Policy Is Failing the Country's Poor," *The Atlantic,* June 24, 2015, https://www.theatlantic.com/business /archive/2015/06/section-8-is-failing/396650.

27. National Low Income Housing Coalition (NLIHC), *The Long Wait for a Home* 6, no. 1 (Fall 2016), http://nlihc.org/sites/default/files/HousingSpotlight_6-1 .pdf; Corianne Payton Scally et al., "The Case for More, Not Less: Shortfalls in Federal Housing Assistance and Gaps in Evidence for Proposed Policy Changes," Urban Institute, Jan. 2018, https://www.urban.org/sites/default/files/publication/95616/case _for_more_not_less.pdf.

28. NLIHC, *The Long Wait for a Home;* Desmond, "Engine of American Inequality." Information about Section 8 availability in Providence appears on the website for the Providence Housing Authority, https://affordablehousingonline.com /housing-authority/Rhode-Island/Providence-Housing-Authority/RI001.

29. For a primer on LIHTC, see, e.g., Mark Keightley, "An Introduction to the Low-Income Housing Tax Credit," Congressional Research Service, Mar. 28, 2018, https://fas.org/sgp/crs/misc/RS22389.pdf; Kirk McClure, "What Should Be the Future of the Low-Income Housing Tax Credit Program?" *Housing Policy Debate* 29, no. 1 (2019): 65–81.

30. The federal HOME Investment Partnerships Program provides a very small grant to states and eligible municipalities that can be put to a variety of uses, including the construction of affordable homes. Purchasers must have incomes at or below 80 percent of the median income for the area. The federal allocation for the HOME Program was $1.25 billion in 2019, less than the amount allocated when the program began in 1992, and provides each participating jurisdiction only a pittance for building affordable private homes. In 2012 the median state grant was about $6 million, and the median locality grant was about $580,000. See "The Home Program: Home Investment Partnerships," Department of Housing and Urban Development, n.d., https://www.hud.gov/hudprograms/home-program; Allison Bell, "2019 Bill Largely Sustains 2018 HUD Funding Gains," Center on Budget and Policy Priorities, Feb. 15, 2019, https://www.cbpp.org/blog/2019-bill-largely-sustains-2018-hud-funding-gains; Katie Jones, "An Overview of the HOME Investment Partnerships Program," Congressional Research Service, Sept. 11, 2014, https://fas.org/sgp/crs/misc/R40118.pdf.

31. I am grateful to the staff of ONE Neighborhood for confirming the facts in this paragraph.

32. "Beneficial Impacts of Homeownership: A Research Summary," Habitat for Humanity, Apr. 2016, http://www.habitatbuilds.com/wp-content/uploads/2016/04/Benefits-of-Homeownership-Research-Summary.pdf.

33. John Emmeus Davis, "Shared Equity Homeownership," National Housing Institute, 2006, https://shelterforce.org/wp-content/uploads/2008/04/SharedEquity Home.pdf.

34. See, e.g., James C. Fraser, Robert J. Chaskin, and Joshua T. Bazuin, "Making Mixed-Income Neighborhoods Work for Low-Income Households," *Cityscape* 15, no. 2 (2013): 87. ("The private sector may choose to participate in developing affordable housing, but it rarely does so unless the state provides deep discounts. Indeed, the number of vacant housing units in the United States would easily house most U.S. citizens, but citizenship rights to housing are eschewed by the private sector and not supported by the U.S. government.")

35. "Income Limits," Department of Housing and Urban Development, Office of Policy Development and Research, https://www.huduser.gov/portal/datasets/il.html; Kirk McClure, "What Should Be the Future of the Low-Income Housing Tax Credit Program?" *Housing Policy Debate* 29, no. 1 (2019): 67 ("The program does not set a floor on tenant household income, but as a practical matter an extremely low-income household, defined by HUD as a household with income below 30% of AMI, cannot afford LIHTC rents as they are set assuming that the tenant household has an income at 50% or 60% of AMI").

36. United States Census, American Community Survey 2018 (5-Year Estimates), Census Tract 19, Providence, Rhode Island, Median Household Income.

37. See, e.g., McClure, "What Should Be the Future," 67 ("The LIHTC program cannot, by itself, assist households with extremely low income"); Lucas Kirkpatrick, "The Two 'Logics' of Community Development: Neighborhoods, Markets, and Community Development Corporations," *Politics and Society* 35 (June 2007): 329–359.

38. Rebecca Diamond and Tim McQuade, "Who Wants Affordable Housing in Their Backyard? An Equilibrium Analysis of Low-Income Property Development," *Journal of Political Economy* 127, no. 3 (June 2019): 1063–1117, https://doi.org/10.1086/701354 ("Investing in affordable housing in low income and high minority areas ... will then spark in-migration of high income and a more racially diverse set of residents"); see also, e.g., Ingrid G. Ellen, Keren M. Horn, and Katherine M. O'Regan, "Poverty Concentration and the Low Income Housing Tax Credit: Effects of Siting and Tenant Composition," *Journal of Housing Economics* 34 (2016): 49–59, https://doi.org/10.1016/j.jhe.2016.08.001 ("The few studies that examine the characteristics of

qualifying LIHTC tenants find that their incomes are higher on average than the incomes of tenants living in other forms of federally-subsidized housing," and "LIHTC developments can make distressed neighborhoods more appealing to higher-income households through removing blight, building attractive new housing, repopulating a community, and/or inviting other investment and improvements"); Gary Bridge, Tim Butler, and Loretta Lees, eds., *Mixed Communities: Gentrification by Stealth?* (Bristol, U.K.: Policy Press, 2012).

39. Diamond and McQuade, "Who Wants Affordable Housing?"

40. Jacqueline Tempera, "Report: 'Major Electrical Problems' at Site of Fatal Providence Fire," *Providence Journal,* Jan. 9, 2018.

41. Patrick Anderson, "9 Injured, 1 Missing as Fire Destroys 3 Olneyville Buildings," *Providence Journal,* Jan. 6, 2018; Jacqueline Tempera, "Providence Building Ordered Condemned Just before Fatal Fire," *Providence Journal,* Jan. 8, 2018; Katie Davis, "Providence Fire Department Releases New Photos of Fatal Fire," *NBC 10 News,* Jan. 9, 2018; Tempera, "Report"; Elizabeth Harrison and John Bender, "It Was a Two-Family Dwelling, with More Than Twenty Occupants. Then It Went Up in Flames," WPRI, Jan. 9, 2018; Sarah Doroin, "Providence Police ID Woman Pulled from Bowdoin Street Fire," WPRI, Jan. 10, 2018.

42. Neil Remiesiewicz and Julianne Lima, "Four Homes Damaged, Several Injuries Reported in Providence Fire," WPRI, Jan. 6, 2018; Nancy Krause, "Body Pulled from Burned-Down Olneyville House," WPRI, Jan. 8, 2018; Tim White, "Inspector Recommended Building Be Condemned Days before Massive Fire," WPRI, Jan. 8, 2018; Paul Doughty, Providence Fire Department arson investigator, "Report of the Arson Investigation into the Cause of the Fire January 6, 2018, at 110 Bowdoin Street."

43. This was not the first time in recent memory that fire took an innocent life on Bowdoin Street. Almost five years to the day before Lucy Feliciano was killed in the fire at 110 Bowdoin, a fire destroyed the multifamily apartment down the street at 88 Bowdoin, killing a ten-month-old child. *NBC 10 News,* "Child Dies in Providence Fire," WJAR, Dec. 22, 2012, https://turnto10.com/archive/child-dies-in-providence-fire; "NBC 10 I-Team: House Destroyed in Fatal Fire Was Being Condemned," WJAR, Jan. 9, 2018, https://turnto10 com/i-team/body-found-in-remains-of-providence-house.

Chapter Five. You Can't Build a House with Handcuffs

1. Lisa M. Sontag-Padilla, Lynette Staplefoote, and Kristy Gonzalez Morganti, "Challenges and Promising Practices of Financial Sustainability in Nonprofit Organizations," in *Financial Sustainability for Nonprofit Organizations: A Review of the Lit-*

erature (Santa Monica, Calif.: RAND Corporation, 2012), 7–19, http://www.jstor
.org/stable/10.7249/j.ctt5hhvjg.8.

2. Isabella retired from the Providence Police Department in 2019 and became
the director of Child Protective Services, a division of the Rhode Island Department
of Children, Youth and Families. In 2020, he was tapped to become the chief of police in Seekonk, Massachusetts. Tom Mooney, "Providence Police Veteran to Lead
DCYF Investigative Unit," *Providence Journal,* July 16, 2019, https://www.provide
ncejournal.com/news/20190716/providence-police-veteran-to-lead-dcyf-investiga
tive-unit; Joe Siegel, "Seekonk Police Chief in Limbo," *Reporter Today,* Oct. 24, 2020,
https://reportertoday.com/stories/seekonk-police-chief-in-limbo,32964?.

3. Even well into the twenty-first century, most urban police departments did
not reflect the racial or ethnic makeup of the city's population. As of 2013, for instance, the Providence Police Department was nearly 79 percent white but only 10
percent Latino, though the city was 40 percent Latino and only 37 percent white.
See "Diversity on the Force: Where Police Don't Mirror Communities," *Governing,*
Sept. 2015, http://images.centerdigitaled.com/documents/policediversityreport.pdf.

4. Simone Weischelbaum and Beth Schwartzapfel, "When Warriors Put on the
Badge," Marshall Project, Mar. 30, 2017, https://www.themarshallproject.org/2017
/03/30/when-warriors-put-on-the-badge. For a discussion of the paramilitary organization in a police department, see Jerome Skolnick and James J. Fyfe, *Above the Law:
Police and the Excessive Use of Force* (New York: Free Press, 1994), 117–118.

5. Seth Stoughton, "Law Enforcement's 'Warrior' Problem," *Harvard Law Review*
128, no. 6 (Apr. 10, 2015), https://harvardlawreview.org/2015/04/law-enforcements
-warrior-problem.

6. Seth Stoughton, "Principled Policing: Warrior Cops and Guardian Officers,"
Wake Forest Law Review 51 (2016): 652.

7. John Bennett, "How Command Presence Affects Your Survival," *PoliceOne,*
Oct. 7, 2010, https://www.policeone.com/Officer-Safety/articles/2748139-How-com
mand-presence-affects-your-survival. Notably, Bennett's admonition to officers para-
phrases a famous quote by General James "Mad Dog" Mattis, the former marine
appointed by President Trump to be his first secretary of defense. According to the
journalist Thomas Ricks, Mattis used to tell his troops, "Be polite, be professional,
but have a plan to kill everybody you meet." Thomas Ricks, "Fiasco," *Armed Forces
Journal,* Aug. 1, 2006, http://armedforcesjournal.com/fiasco. In a meeting with Iraqi
military leaders, Mattis reportedly said, "I come in peace. I didn't bring artillery. But
I'm pleading with you, with tears in my eyes: If you fuck with me, I'll kill you all."

8. Skolnick and Fyfe, *Above the Law,* 111, quoting Paul Chevigny, *Police Power: Police Abuses in New York City* (New York: Vintage, 1969).

9. Mitch Smith and Timothy Williams, "Minnesota Police Officer's 'Bulletproof Warrior' Training Is Questioned," *New York Times,* July 4, 2016, https://www.nytimes.com/2016/07/15/us/minnesota-police-officers-bulletproof-warrior-training-is-questioned.html.

10. *Patriot Act,* season 4, episode 6, "The Broken Policing System," directed by Richard A. Pruess, written by Hasan Minhaj, Prashanth Venkataramanujam, and Scott Vrooman, aired Sept. 2019, on Netflix. For a particularly colorful account of the warrior mind-set as imagined by David Grossman, see Steve Featherstone, "Professor Carnage," *New Republic,* Apr. 17, 2017, https://newrepublic.com/article/141675/professor-carnage-dave-grossman-police-warrior-philosophy.

11. *Patriot Act,* "The Broken Policing System."

12. Skolnick and Fyfe, *Above the Law,* 101–103, quoting Paul Chevigny, *Police Power: Police Abuses in New York City* (New York: Vintage, 1969).

13. Much has been written about the Bland case. The dash cam video of the increasingly hostile altercation between Bland and Encinia is widely available on YouTube. See, e.g., https://www.nytimes.com/video/us/100000003813646/police-video-shows-sandra-blands-arrest.html.

14. Doug Wyllie, "How Strong Command Presence Can Quickly Resolve Dynamic Incidents," *PoliceOne,* Mar. 10, 2017, https://www.policeone.com/Officer-Safety/articles/299987006-How-strong-command-presence-can-quickly-resolve-dynamic-incidents.

15. Skolnick and Fyfe, *Above the Law,* 116.

16. Skolnick and Fyfe, *Above the Law,* 211–216; "Shielded from Justice: Police Brutality and Accountability in the United States," Human Rights Watch, June 1998, https://www.hrw.org/legacy/reports98/police/index.htm.

17. Amanda Milkovits, "Former Providence Police Chief Urbano Prignano Dies," *Providence Journal,* Nov. 22, 2017, https://www.providencejournal.com/news/20171122/former-providence-police-chief-urbano-prignano-dies.

18. Amanda Milkovits, "'Buddy's Boys': Report Documents Long History of Collusion, Influence with Providence Police," *Newport Daily News,* updated Oct. 6, 2015, https://www.newportri.com/article/20141018/NEWS/310189994.

19. Milkovits, "Buddy's Boys"; Stanton, *The Prince of Providence.*

20. For a lively and thorough discussion of Cianci's trial, sentencing, and final months as a free man, see Stanton, *The Prince of Providence,* chaps. 14–15.

21. Quoted in Rob Gurwitt, "Bratton's Brigade," *Governing*, Aug. 1, 2007, https://www.governing.com/topics/public-justice-safety/Brattons-Brigade.html.

22. Herold is now the chief of police in Boulder, Colorado.

23. John Gramlich, "Most Violent and Property Crimes in the U.S. Go Unsolved," Pew Research Center, Mar. 1, 2017, https://www.pewresearch.org/fact-tank/2017/03/01/most-violent-and-property-crimes-in-the-u-s-go-unsolved/; German Lopez, "There's Nearly a 40 Percent Chance You'll Get Away with Murder in America," Vox, Sept. 24, 2018, https://www.vox.com/2018/9/24/17896034/murder-crime-clearance-fbi-report.

24. Gurwitt, "Bratton's Brigade."

25. Stephen Beale, "Breaking News: Esserman Resigns," *GoLocalProv*, June 22, 2011, https://www.golocalprov.com/news/breaking-news-esserman-resigns.

26. "New Haven Police Chief Dean Esserman Resigns Effective Sept. 2," *New Haven Register*, Sept. 6, 2016, https://www.nhregister.com/connecticut/article/New-Haven-Police-Chief-Dean-Esserman-resigns-11327411.php.

27. For a discussion of the malleability of the label *community policing*, see, e.g., Jihong Zhao, Nicholas P. Lovrich, and Quint Thurman, "The Status of Community Policing in American Cities: Facilitators and Impediments Revisited," *Policing: An International Journal* 22, no. 1 (1999): 74–92. ("Theoretically, community policing represents a new philosophy of doing police work; however, at a more practical level of actual policing practices, a wide variety of activities can be attributed to community policing—ranging from genuine community-police partnerships on one end to a zero-tolerance for specific crimes on the other end of the spectrum. For example, some researchers have shown that a substantial proportion of officers were dispensing a 'tough' and traditional law enforcement set of activities, but doing so under the banner of community policing.")

28. For an account of hostility among line officers toward the officers assigned to community policing units, see, e.g., Susan Sadd and Randolph Grinc, "Innovative Neighborhood Oriented Policing: An Evaluation of Community Policing Programs in Eight Cities," in *The Challenge of Community Policing: Testing the Promises*, ed. Dennis P. Rosenbaum (Thousand Oaks, Calif.: SAGE, 1994), 27–52.

Chapter Six. "What Was Done, Was Done Right"

1. Anthony A. Braga, Andrew V. Papachristos, and David M. Hureau, "The Concentration and Stability of Gun Violence at Micro Places in Boston, 1980–2008," *Journal of Qualitative Criminology* 26, no. 1 (Mar. 2010): 33–53.

2. Braga, Papachristos and Hureau, "Concentration and Stability of Gun Violence."

3. David Weisburd, John E. Eck, Anthony A. Braga, Cody W. Telep, Breanne Cave, et al., *Place Matters: Criminology for the Twenty-First Century* (New York: Cambridge University Press, 2016), 18–22.

4. Bill Geller and Lisa Belsky, "A Study of Providence's Olneyville Neighborhood," in *Building Our Way Out of Crime: The Transformative Power of Police-Community Developer Partnerships* (Glenview, Ill.: Geller & Associates, 2012), 5–6, http://www.olneyville.org/Geller-Belsky-case-study.pdf.

5. Weisburd et al., *Place Matters,* chap. 3.

6. For more examples of place-based problem-solving, see "Arizona State University Center for Problem-Oriented Policing," ASU Watts College of Public Service and Community Solutions, https://popcenter.asu.edu.

7. Geller and Belsky, "A Study of Providence," 21.

8. Geller and Belsky, "A Study of Providence," 21.

9. Geller and Belsky, "A Study of Providence," 25.

10. Fernandes has since been promoted to captain.

11. This is not to say that residents are indifferent to violence. Quite the contrary, when violence threatens residents who are not involved in criminal activity, they are exceedingly concerned and often implore the police to increase patrols. In October 2020, as I was putting the last touches on this book, Captain Fernandes met with about fifty residents in Elmhurst, another Providence neighborhood, where gang violence and gunplay were beginning to threaten innocent bystanders. Though some activists in Providence have called to defund the police department, several residents at this meeting "said they wanted more officers, not fewer." Brian Amaral, "Crime Concerns: Residents, Police Discuss Increase in Some Crimes in Providence," *Providence Journal,* Oct. 18, 2020, https://www.providencejournal.com/story/news/2020/10/18/crime-concerns-residents-police-discuss-increase-in-some-crimes-in-providence/42854943.

12. Monica Bell, "The Community in Criminal Justice: Subordination, Consumption, Resistance, and Transformation," *Du Bois Review* 16, no. 1 (2019): 204.

13. William H. Sousa and George L. Kelling, "Of 'Broken Windows,' Criminology, and Criminal Justice," in *Police Innovation: Contrasting Perspectives,* ed. David Weisburd and Anthony A. Braga (New York: Cambridge University Press, 2006), 77.

14. James Q. Wilson, "Just Take Away Their Guns," *New York Times Magazine,* March 20, 1994, https://www.nytimes.com/1994/03/20/magazine/just-take-away

-their-guns.html; Jeffrey Bellin, "The Inverse Relationship between the Constitutionality and Effectiveness of New City 'Stop and Frisk,'" *Boston University Law Review* 94 (2014): 1550.

15. Terry v. Ohio, 392 U.S. 1 (1968).

16. Wilson, "Just Take Away Their Guns."

17. See "Annual Stop-and-Frisk Numbers," New York Civil Liberties Union, https://www.nyclu.org/en/stop-and-frisk-data.

18. Ross Tuttle and Quinn Rose Schneider, "Stopped-and-Frisked: 'For Being a F**king Mutt,'" *The Nation*, Oct. 8, 2012, https://www.thenation.com/article/stopped-and-frisked-being-fking-mutt-video. The article contains a link to the two-minute audio of the stop.

19. As of November 2020, the footage was at "NYPD Stop & Frisk—Spot Check," Allthingsharlem, posted to YouTube, Oct. 17, 2010, https://www.youtube.com/watch?v=9S2FnE7rmVs.

20. Tuttle and Schneider, "Stopped-and-Frisked."

21. Joseph Goldstein and Ashley Southall, "I Got Tired of Hunting Black and Hispanic People," *New York Times*, Dec. 9, 2019, https://www.nytimes.com/2019/12/06/nyregion/nyc-police-subway-racial-profiling.html.

22. Juan Del Toro et al., "The Criminogenic and Psychological Effects of Police Stops on Adolescent Black and Latino Boys," *Proceedings of the National Academy of Sciences* 116, no. 17 (Apr. 23, 2019): 8261–8268.

23. Richard Rosenfeld and Robert Fornango, "The Relationship between Crime and Stop, Question, and Frisk Rates in New York City Neighborhoods," *Justice Quarterly* 34, no. 6 (2017): 931–951.

24. Bellin, "The Inverse Relationship," 1550. ("The NYPD brought a taste of prison to the street, putting thousands of innocent New Yorkers through the types of invasive scrutiny one would expect in confinement.")

25. Michael D. White and Henry F. Fradella, *Stop and Frisk: The Use and Abuse of a Controversial Policing Tactic* (New York: NYU Press, 2016), 93; Al Baker and Joseph Goldstein, "2 Opinions on Stop-and-Frisk Report," *New York Times*, May 9, 2012; Eric T. Schneiderman, "A Report on Arrests Arising from the New York City Police Department's Stop-and-Frisk Practices," New York State Office of the Attorney General, Nov. 2013, 1, https://ag.ny.gov/pdfs/OAG_REPORT_ON_SQF_PRACTICES_NOV_2013.pdf.

26. James Q. Wilson and George Kelling, "Broken Windows," *The Atlantic*, Mar. 1982, https://www.theatlantic.com/magazine/archive/1982/03/broken-windows/304465.

27. See, e.g., George L. Kelling, "Reclaiming the Subway," *City Journal*, Winter 1991, https://www.city-journal.org/html/reclaiming-subway-12770.html.

28. National Academies of Sciences, Engineering, and Medicine, *Proactive Policing: Effects on Crime and Communities* (Washington, D.C.: National Academies Press, 2018), https://doi.org/10.17226/24928.

29. Elaine B. Sharp, "Politics, Economics, and Urban Policing: The Postindustrial City Thesis and Rival Explanations of Heightened Order Maintenance Policing," *Urban Affairs Review* 50, no. 3 (2014): 340–365, https://doi.org/10.1177/107808741 3490397.

30. Timothy A. Gibson, *Securing the Spectacular City: The Politics of Revitalization and Homelessness in Downtown Seattle* (Lanham, Md.: Lexington Books, 2004).

31. Ayobami Laniyonu, "Coffee Shops and Street Stops: Policing Practices in Gentrifying Neighborhoods," *Urban Affairs Review* 54, no. 5 (2018): 898–930, https://doi.org/10.1177/1078087416689728.

32. Mike Davis, "Fortress Los Angeles: The Militarization of Urban Space," in *Variations on a Theme Park: Scenes from the New American City and the End of Public Space*, ed. M. Sorkin (New York: Hill & Wang, 1992), 154–180; Robert J. Chaskin and Mark L. Joseph, "Contested Space: Design Principles and Regulatory Regimes in Mixed-Income Communities in Chicago," *Annals of the American Academy of Political and Social Science* 660, no. 1 (2015): 136–154, https://doi.org/10.1177/0002716215576113; Lam Thuy Vo, "They Played Dominoes outside Their Apartment for Decades. Then the White People Moved In and Police Started Showing Up," *Buzzfeed News*, June 29, 2018, https://www.buzzfeednews.com/article/lamvo/gentrification-complaints-311 -new-york.

33. Similar collaborations have come together across the country, and code enforcement is now a well-known part of the neighborhood revitalization process. See, e.g., Joe Schilling and Elizabeth Schilling, "Leveraging Code Enforcement for Neighborhood Safety: Insights for Community Developers," LISC, July 2007, https://www .communityprogress.net/filebin/pdf/nvpc_trnsfr/Schilling_LeveragingCodeEnf.pdf.

34. Geller and Belsky, "A Study of Providence," 21–24.

35. Geller and Belsky, "A Study of Providence," 21–24.

36. Geller and Belsky, "A Study of Providence," 24–26.

37. See William J. Bratton and Jon Murad, "Precision Policing," *City Journal*, Summer 2018, https://www.city-journal.org/html/precision-policing-16033.html.

38. Geller and Belsky, "A Study of Providence," 41–48. Improving the area around Riverside Park and Aleppo Street has not led to more crime elsewhere in the neighborhood. On the contrary, crime has continued to fall in Olneyville and District 5.

Between 2013 and December 18, 2018, homicides dropped by two-thirds, aggravated assaults and burglary by more than half, and forcible sex offenses and weapons offenses by more than 20 percent. See Providence Police Department, "Part I—Crime Comparison Report, District 5, Five Year Weighted Average through Dec. 18, 2018," https://www.providenceri.gov/wp-content/uploads/2018/12/20181216.pdf (the last available report of 2018 provides data through December 18, rather than year-end).

39. Amanda Milkovits, "R.I. Groups Serving Male and Female Prostitutes to Merge," *Providence Journal,* Mar. 26, 2016, https://www.providencejournal.com/arti cle/20160326/news/160329408.

40. Dean Isabella, "What Was Done, Was Done Right," Rhode Island LISC, Mar. 22, 2017, http://www.lisc.org/our-stories/story/what-was-done-was-done-right. In 2011, Isabella received the inaugural L. Anthony Sutin Civic Imagination Award from the Department of Justice. This award is given annually to recognize "the efforts of collaborative partnerships within the community. This award is bestowed on a team of law enforcement and community members whose innovative civic interactions have transformed public safety and contributed to violent crime reduction in their community." See L. Anthony Sutin Civic Imagination Award, https://cops .usdoj.gov/sutinaward.

Chapter Seven. "It's an Oasis"

1. "Providence Public School District: A Review," Johns Hopkins School of Education, Institute for Education Policy, June 2019, https://edpolicy.education.jhu .edu/wp-content/uploads/2019/11/PPSD-REVISED-FINAL.pdf.

2. Linda Borg, "5 Things to Know about the Scathing Report on Providence Schools," *Providence Journal,* June 27, 2019, https://www.providencejournal.com/news /20190625/5-things-to-know-about-scathing-report-on-providence-schools.

3. Steph Machado and Eli Sherman, "12 Things to Know about the Devastating New Providence Schools Report," WPRI.com, June 25, 2019, https://www.wpri.com /news/local-news/providence/12-things-to-know-about-the-devastating-new-provi dence-schools-report.

4. Valerie Strauss, "Blistering Report Details Abject Dysfunction and Dangerous Schools in Providence, R.I.," *Washington Post,* June 26, 2019, https://www.washing tonpost.com/education/2019/06/26/blistering-report-details-abject-dysfunction-dan gerous-schools-providence-ri.

5. "An Education Horror Show," *Wall Street Journal,* July 7, 2019.

6. "Providence Public School District: A Review," 77; Livia Gimenez, "D'Abate, Swearer Partnership Praised for Success," *Brown Daily Herald,* Oct. 29, 2019,

https://www.browndailyherald.com/2019/10/29/dabate-swearer-partnership-praised
-success. Brent Kerman also confirmed to me that the unnamed school was D'Abate
Elementary.

7. Stephanie Levin and Kathryn Bradley, "Understanding Principal Turnover:
A Review of the Research," National Association of Secondary School Principals,
Learning Policy Institute, 2017, https://learningpolicyinstitute.org/sites/default/files
/product-files/NASSP_LPI_Principal_Turnover_Research_Review_REPORT.pdf;
Soheyla Taie and Rebecca Goldring, "Characteristics of Public Elementary and Sec-
ondary School Principals in the United States: Results from the 2015–16 National
Teacher and Principal Survey," U.S. Department of Education (Washington, D.C.:
National Center for Education Statistics), https://nces.ed.gov/pubsearch/pubsinfo
.asp?pubid=2017070.

8. David Holmstrom, "One Stop Help for Urban Children, Parents," *Christian
Science Monitor,* Apr. 11, 1994, https://www.csmonitor.com/1994/0411/11101.html.

9. "Rhode Island Elementary School Rankings," SchoolDigger, https://www
.schooldigger.com/go/RI/schoolrank.aspx; "William D'Abate Elementary School,"
Public School Review, https://www.publicschoolreview.com/william-d-abate-elemen
tary-school-profile; "William D'Abate Elementary School," Niche, https://www.niche
.com/k12/william-dabate-elementary-school-providence-ri.

10. The surveys for every school are available at the website for the Rhode Island
Department of Education, https://www.ride.ri.gov. The report card for D'Abate Elementary is at https://reportcard.ride.ri.gov/201819/SchoolSnapshot?SchCode=28153
&DistCode=28. The site permits comparisons to averages both citywide and statewide.

11. The report card for the Cumberland Community School in Cumberland,
Rhode Island, is at https://reportcard.ride.ri.gov/201819/SchoolSnapshot?SchCode
=08109&DistCode=08. Information on the demographics of D'Abate and the
Cumberland Community School, including racial distribution, median household
income, and percent entitled to free or discounted lunch, can be found at a number
of sites, including www.niche.com and www.schooldigger.com.

12. National Commission on Excellence in Education, *A Nation at Risk: The Im-
perative for Educational Reform,* Apr. 1983, https://files.eric.ed.gov/fulltext/ED226006
.pdf.

13. Christine Patterson, "Measuring the Lasting Impact of 'A Nation at Risk,'"
Walton Family Foundation, Apr. 10, 2018, quoting Bruno Manno, https://www.wal
tonfamilyfoundation.org/stories/k-12-education/measuring-the-lasting-impact-of
-a-nation-at-risk.

14. Carnegie Foundation for the Advancement of Teaching, *An Imperiled Generation: Saving Urban Schools* (Princeton, N.J.: Carnegie Foundation, 1988), xii–xiii, https://files.eric.ed.gov/fulltext/ED293940.pdf. Apocalyptic warnings are a staple of these reports. See, e.g., "A Stagnant Nation: Why American Students Are *Still* at Risk," ED in '08, Strong American Schools Project, Apr. 2008, https://www.issuelab .org/resources/11189/11189.pdf ("Time is running out on America's opportunity to enact a robust national education reform agenda.... Our students deserve better, and our nation's economic security is at greater risk now than ever before").

15. "Imagine: Providence Blueprint for Education," report of the PROBE Commission, May 1993, https://files.eric.ed.gov/fulltext/ED365762.pdf.

16. See, e.g., Thomas Petrilli, "Charters Can Do What's Best for Students Who Care," *New York Times,* Jan. 19, 2016, https://www.nytimes.com/roomfordebate /2014/12/10/are-charter-schools-cherry-picking-students/charters-can-do-whats-best -for-students-who-care.

17. See Diane Ravitz, *The Death and Life of the Great American School: How Testing and Choice Are Undermining Education* (New York: Basic Books, 2016), chaps. 6, 8; Jeffrey R. Henig, *Rethinking School Choice: The Limits of the Market Metaphor* (Princeton, N.J.: Princeton University Press, 1994).

18. Many scholars have traced the arc of neoliberal education reform. See, e.g., Pauline Lipman, *The New Political Economy of Urban Education: Neoliberalism, Race, and the Right to the City* (New York: Taylor & Francis, 2011). In addition to charters and testing, vouchers have become another part of the education reform agenda. In some states, parents are given a voucher for each school-age child equal to the amount of money spent by the state, per pupil, on public education. As with Section 8 vouchers in housing, parents are then free to use that education voucher toward the cost of tuition at private schools, most of which are affiliated with a particular religion. Public funding for private religious schools is important to some on the religious right (see, e.g., Michael W. Apple, *Educating the "Right" Way: Markets, Standards, God, and Inequality* [New York: Taylor & Francis, 2006]), but it is not a prominent part of the neoliberal agenda, and only eighteen states make vouchers available. Rhode Island is not among them. Instead it allows businesses to make charitable donations to organizations that award scholarships to private schools, and in exchange the business receives a generous tax credit. This arrangement is more common than vouchers and is available in twenty-three states. See "Fast Facts on School Choice," EdChoice, https://www.edchoice.org/home/fast-facts-resources.

19. Ravitz, *Death and Life,* 158–161. Notably, Shanker withdrew his support for charters after he saw what they became (159).

20. Kevin Carey, "The Demise of the Great Education Saviors," *Washington Post Magazine,* Mar. 18, 2020, https://www.washingtonpost.com/magazine/2020/03/18/charter-schools-testing-were-supposed-save-american-education-now-theyve-run-out-political-steam-what-went-wrong/?arc404=true.

21. See, e.g., David C. Berliner, "Poverty and Potential: Out-of-School Factors and School Success," Great Lakes Center for Education Research and Practice, Mar. 2009, https://greatlakescenter.org/docs/Policy_Briefs/Berliner_NonSchool.pdf; Susan B. Neuman, *Changing the Odds for Children at Risk: Seven Essential Principles of Educational Programs That Break the Cycle of Poverty* (New York: Teachers College Press, 2009); Matthew DiCarlo, "Teachers Matter, but So Do Words," *Shanker Blog,* Albert Shanker Institute, July 14, 2010, https://www.shankerinstitute.org/blog/teachers-matter-so-do-words.

22. Kenneth Leithwood, Karen Seashore Louis, Stephen Anderson, and Kyla Wahlstrom, "Review of Research: How Leadership Influences Student Learning," Center for Applied Research and Educational Improvement, 70, https://www.wallacefoundation.org/knowledge-center/documents/how-leadership-influences-student-learning.pdf. ("Of all the factors that contribute to what students learn at school, present evidence led us to the conclusion that leadership is second in strength only to classroom instruction.")

23. Douglas B. Downey et al., "Are Schools the Great Equalizer? Cognitive Inequality during the Summer Months and the School Year," *American Sociological Review* 69, no. 5 (2004): 613–635.

24. Susan B. Neuman, "Use the Science of What Works to Change the Odds for Children at Risk," *Phi Delta Kappan,* Apr. 2009, 582–587, https://journals.sagepub.com/doi/10.1177/003172170909000812.

25. *A Nation at Risk,* appendix A, 39.

26. James Crawford, "A Diminished Vision of Civil Rights," *Education Week,* June 5, 2007, https://www.edweek.org/ew/articles/2007/06/06/39crawford.h26.html; Berliner, "Poverty and Potential," 6–7.

27. "Why the Achievement Gap Persists," *New York Times,* Dec. 8, 2006.

28. Sean F. Reardon, "The Widening Income Achievement Gap," *Faces of Poverty* 70, no. 8 (May 2013): 10–16, http://www.ascd.org/publications/educational-leadership/may13/vol70/num08/The-Widening-Income-Achievement-Gap.aspx.

29. Joseph Stiglitz, "Three Decades of Neoliberal Policies Have Decimated the Middle Class, Our Economy, and Our Democracy," *MarketWatch,* May 13, 2019, https://www.marketwatch.com/story/three-decades-of-neoliberal-policies-have-decimated-the-middle-class-our-economy-and-our-democracy-2019-05-13;

Christopher Ingraham, "The Richest 1 Percent Now Owns More of the Country's Wealth Than at Any Time in the Past 50 Years," *Washington Post,* Dec. 6, 2017, https://www.washingtonpost.com/news/wonk/wp/2017/12/06/the-richest-1-percent-now-owns-more-of-the-countrys-wealth-than-at-any-time-in-the-past-50-years.

30. David Leonhardt, "The Rich Really Do Pay Lower Taxes Than You," *New York Times,* Oct. 6, 2019, https://www.nytimes.com/interactive/2019/10/06/opinion/income-tax-rate-wealthy.html.

31. See Rebecca David and Kevin Hesla, "Estimated Public Charter School Enrollment, 2017–2018," National Alliance for Public Charter Schools, 2018, https://www.publiccharters.org/sites/default/files/documents/2018-03/FINAL%20Estimated%20Public%20Charter%20School%20Enrollment%2C%202017-18.pdf; Digest of Education Statistics, National Center for Education Statistics, table 203.10, "Enrollment in Public Elementary and Secondary Schools," https://nces.ed.gov/programs/digest/d18/tables/dt18_203.10.asp.

32. See "50 State Comparison," Education Commission of the States, Jan. 2020, http://ecs.force.com/mbdata/MBQuestNB2C?rep=CS2014.

33. Rhode Island Department of Education, "The State of Rhode Island's Charter Schools," Apr. 2014, https://www.ride.ri.gov/Portals/0/Uploads/Documents/Students-and-Families-Great-Schools/Charter-Schools/State_of_RI_Charter_Public_Schools_FINAL.pdf.

34. Ravitz, *Death and Life,* 129–131.

35. William J. Mathis and Tina Trujillo, "Lessons from NCLB for the Every Student Succeeds Act," National Education Policy Center, Nov. 2016, http://nepc.colorado.edu/publication/lessons-from-NCLB.

36. "National Charter School Study 2013," Stanford Center for Research on Education Outcomes, 2013, https://credo.stanford.edu/sites/g/files/sbiybj6481/f/ncss_2013_final_draft.pdf. One review of the Stanford study suggests even this slight improvement may overstate the case. See Andrew Maul and Abby McClelland, "Review of National Charter School Study 2013," National Education Policy Center, July 2013, https://nepc.colorado.edu/thinktank/review-credo-2013.

37. Carey, "Demise of the Great Education Saviors." See also Gary Miron, William Mathis, and Kevin Welner, "Review of Separating Fact and Fiction: What You Need to Know," National Education Policy Center, Feb. 2015, 10, https://nepc.colorado.edu/thinktank/review-separating-fact-and-fiction ("After controlling for student demographics, charter schools show test-score results at levels that are not meaningfully better or worse than district schools").

38. "Do Charter Schools and School Vouchers 'Hurt' Public Education?" Network for Public Education, n.d., collecting case studies, https://networkforpubliceducation .org/wp-content/uploads/2019/01/Do-charter-schools-and-school-vouchers-hurt -public-schools%C6%92.pdf; "Separating Fact and Fiction," 12 ("When children move from public schools to charter schools, the traditional public schools lose money that then goes to the charter schools. Accordingly, in this ideal narrative, charter schools do in fact take money away from traditional public schools").

39. Yongmei Ni and David Arsen, "School Choice Participation Rates: Which Districts Are Pressured?" *Education Policy Analysis Archives* 19, no. 29 (Oct. 2011), http://epaa.asu.edu/ojs/article/view/777; Gary Miron, Jessica L. Urschel, William J. Mathis, and Elana Tornquist, "Education and the Public Interest Center and Education Policy Research Unit," 2010, http://epicpolicy.org/publication/schools-without -diversity.

40. "Separating Fact and Fiction," 8–9.

41. For a comprehensive review of the literature on testing and its application to No Child Left Behind, see Kevin G. Welner and William J. Mathis, "Reauthorization of the Elementary and Secondary Education Act: Time to Move beyond Test-Focused Policies," National Education Policy Center, Feb. 2015, https://nepc.colo rado.edu/sites/default/files/nepc-policymemo-esea.pdf.

42. Paul Tough, *The Years That Matter Most: How College Makes or Breaks Us* (Boston: Houghton Mifflin Harcourt, 2019).

43. Ravitz, *Death and Life*, 142–143.

44. Neal McCluskey, "Has No Child Left Behind Worked?" CATO Institute, Feb. 9, 2015, https://www.cato.org/publications/testimony/has-no-child-left-behind -worked; see also Jaekyung Lee and Todd Reeves, "Revisiting the Impact of NCLB High-Stakes School Accountability, Capacity, and Resources: State NAEP 1990– 2009 Reading and Math Achievement Gaps and Trends," *Educational Evaluation and Policy Analysis* 34, no. 2 (June 2012).

45. L. Sutcher, L. Darling-Hammond, and D. Carver-Thomas, *A Coming Crisis in Teaching? Teacher Supply, Demand, and Shortages in the U.S.* (Palo Alto, Calif.: Learning Policy Institute, 2016), 49–50, https://learningpolicyinstitute.org/sites/default /files/product-files/A_Coming_Crisis_in_Teaching_REPORT.pdf.

46. Linda Borg, "At Forum, Some Providence Parents Plead for State to Take Over Schools," *Providence Journal,* June 27, 2019, https://www.providencejournal .com/news/20190626/at-forum-some-providence-parents-plead-for-state-to-take -over-schools.

47. Linda Borg, "State Takeover of Providence Schools Starts Nov. 1, Will Last 5 Years," *Providence Journal*, Oct. 17, 2019, https://www.providencejournal.com/news /20191015/state-takeover-of-providence-schools-starts-nov-1-will-last-5-years.

48. Adam Harris, "Why Rhode Island's Governor Is Taking Over Providence's Public Schools," *The Atlantic*, Oct. 29, 2019, https://www.theatlantic.com/education /archive/2019/10/rhode-island-governor-explains-providence-school-takeover/601003.

49. For more on neighborhood schools, see Coalition for Community Schools, www.communityschools.org.

Chapter Eight. "They're Pimping Us"

1. Dean Esserman, who reshaped the Providence Police Department, Hugh Clements, who designed the police strategy at Aleppo Street and replaced Esserman as chief, and Dean Isabella, who implemented Clements's plan, are also white. So is Brent Kerman, the longtime principal at D'Abate Elementary.

2. See, e.g., Anastasia Reesa Tomkin, "How White People Conquered the Non-profit Industry," Medium.com, Apr. 28, 2020, https://medium.com/an-injustice/how -white-people-conquered-the-non-profit-industry-c1221cd93a83.

3. Brian D. Goldstein, *The Roots of Urban Renaissance: Gentrification and the Struggle Over Harlem* (Cambridge, Mass.: Harvard University Press, 2017), 119; see also Avis C. Vidal, "Rebuilding Communities: A National Study of Urban Community Development Corporations," Community Development Research Center, Graduate School of Management and Urban Policy, New School for Social Research (New York, 1992), 33. For a survey of the history of nonprofit organizations, see David C. Hammack, "Nonprofit Organizations in American History: Research Opportunities and Sources," *American Behavioral Scientist* 45, no. 11 (2020): 1638–1674, https://doi.org/10.1177/0002764202045011004.

4. Goldstein, *Urban Renaissance*, 123.

5. Goldstein, *Urban Renaissance*, 123–125.

6. Goldstein, *Urban Renaissance*, 210; Demetrios Caraley, "Washington Abandons the Cities," *Political Science Quarterly* 107, no. 1 (1992): 1–30.

7. Martha Burt, *Over the Edge: The Growth of Homelessness in the 1980s* (New York: Russell Sage Foundation, 1992); Randy Shilts, *And the Band Played On: Politics, People and the AIDS Epidemic* (New York: Saint Martin's Press, 1987); "The Crack Cocaine Epidemic: Health Consequences and Treatment," Government Accounting Office, Jan. 1991, https://www.gao.gov/assets/90/89031.pdf; Caraley, "Washington Abandons the Cities," 2.

8. See, e.g., Jeff Speck, "Why 12-Foot Traffic Lanes Are Disastrous for Safety

and Must Be Replaced Now" *CityLab,* Oct. 6, 2014, https://www.citylab.com/design /2014/10/why-12-foot-traffic-lanes-are-disastrous-for-safety-and-must-be-replaced -now/381117/; Angie Schmitt, "Compelling Evidence That Wider Lanes Make City Streets More Dangerous," *Streetsblog,* May 27, 2015, https://usa.streetsblog.org/2015 /05/27/compelling-evidence-that-wider-lanes-make-city-streets-more-dangerous.

9. Jeremy Levine, *Constructing Community: Urban Governance, Community Development, and Neighborhood Inequality in Boston* (Princeton, N.J.: Princeton University Press, 2021); see also Vidal, "Rebuilding Communities," 53–62; Steven Rathgeb Smith and Michael Lipsky, *Nonprofits for Hire: The Welfare State in the Age of Contracting* (Cambridge, Mass.: Harvard University Press, 1993).

10. Both Lehrer and Nickerson are white.

11. Though retired, Father Tetrault remains a passionate advocate for the Latino community. See Lauren Clem, "Diocesan Priest to Be Awarded for Evangelization, Loving Service to Hispanic Community," *Rhode Island Catholic,* Mar. 31, 2016, http:// www.thericatholic.com/stories/Diocesan-priest-to-be-awarded-for-evangelization -loving-service-to-Hispanic-community,8078.

12. Catherine E. Wilson, *The Politics of Latino Faith: Religion, Identity, and Urban Community* (New York: NYU Press, 2008), 4–5.

13. Wilson, *The Politics of Latino Faith,* 14–37.

14. See, e.g., Sean Thomas-Breitfeld and Frances Kunreuther, "Race to Lead: Confronting the Nonprofit Racial Leadership Gap," Building Movement Project, 2017, https://racetolead.org/race-to-lead/; Nuri Heckler, "Whiteness and Masculinity in Nonprofit Organizations: Law, Money, and Institutional Race and Gender," *Administrative Theory and Praxis* 41, no. 3 (2019): 266–285, https://doi.org/10.1080 /10841806.2019.1621659.

Chapter Nine. Thanks for Everything (Now Get Out)

1. William Morgan, "Olneyville's Mansion: Atlantic Mills," *GoLocalProv,* August 26, 2016, https://www.golocalprov.com/business/Olneyvilles-Mansion-Atlantic -Mills.

2. Morgan, "Olneyville's Mansion."

3. In 2014, for instance, a writer for another Providence online magazine thought Olneyville might be "ground zero for what's wrong with the U.S. economy." James Kennedy, "Supersize Your Development," *ecoRInews,* Feb. 21, 2014, https:// www.ecori.org/smart-growth/2014/2/21/supersize-your-development.html.

4. Alisha A. Pina, "A Cavalry of Community Groups Rebuilds Olneyville," *Providence Journal,* Mar. 2, 2014, https://www.providencejournal.com.

5. Pina, "Cavalry of Community Groups."

6. Karen A. Davis, "A Rebirth for Olneyville," The Plant Providence, Aug. 31, 2005, http://www.theplantprovidence.com/press/083105.

7. "The Mind 2 Body Fitness Studio," Mind 2 Body Fitness, http://www.mind 2bodyfit.com/Studio.html.

8. "About," Ello Pretty, https://www.ellopretty.com/about.

9. "The Plant," Armory Management Company, http://armorymanagement .com/the-plant.

10. "Location + Hours," Riffraff, http://www.riffraffpvd.com.

11. Sam Sparkes, "These Are the Ten Worst Neighborhoods in Providence for 2019," RoadSnacks.net, Dec. 28, 2018, https://www.roadsnacks.net/worst-neighbor hoods-in-providence-ri.

12. Max Read, "This Is the Williamsburg of Your City: A Map of Hip America," Gawker Magazine, Jan. 29, 2014, https://gawker.com/this-is-the-williamsburg-of -your-city-a-map-of-hip-ame-1460243062.

13. Jennifer Hayes, "Community of Design," The Take, Feb. 27, 2017, https:// thetakemagazine.com/community-of-design. "PVD" refers to Providence.

14. Barry Fain et al., "20 Reasons We Love Providence," East Side Monthly, Sept. 14, 2017, http://eastsidemonthly.com/stories/20-reasons-we-love-providence,24802.

15. Amanda M. Grosvenor, "The New-New Olneyville," Providence Monthly, Oct. 23, 2017, http://providenceonline.com/stories/olneyville-november-providence -monthly,25226.

16. See, e.g., Camryn Rabideau, "The Five Best Neighborhoods in Providence," SquareFoot.com, Apr. 26, 2018 ("artsy, up-and-coming").

17. Chelsey DiCenzo, "Wicked Romantic Must-Do's in RI," Providence Monthly, Feb. 12, 2019, http://providenceonline.com/stories/wicked-romantic-must-dos-in -ri,30258; Robert Isenberg, "Top Shelf," Providence Monthly, Mar. 6, 2019, http:// providenceonline.com/stories/top-shelf,30624.

18. "Stylish Studio in a Historic Mill," https://www.agreatertown.com/provi dence_ri/1br_studio_vacation_rental_in_providence_rhode_island_0003329777.

19. Grosvenor, "New-New."

20. Amanda M. Grosvenor, "Park Yourself Here," Providence Monthly, July 24, 2018, http://providenceonline.com/stories/park-yourself-here,28176.

21. Fay Strongin, "You Don't Have a Problem Until You Do: Revitalization and Gentrification in Providence, Rhode Island" (MA thesis, City Planning, MIT, June 2017), https://dspace.mit.edu/bitstream/handle/1721.1/111259/1003291357-MIT .pdf?sequence=1.

22. U.S. Census Bureau, 2000 Decennial Digest and American Community Surveys, 2018 (5-Year Estimates), Census Tract 19, Providence County, "Educational Attainment of Population over 25."

23. U.S. Census Bureau, American Community Surveys, 2009 and 2018 (5-Year Estimates), Census Tract 19, Providence County, "Gini Index of Inequality"; "Hispanic Latino by Race."

24. For a comprehensive review of the literature, see Miriam Zuk et al., "Gentrification, Displacement and the Role of Public Investment: A Literature Review," in *Federal Reserve Bank of San Francisco* (Federal Reserve Bank of San Francisco, 2015), 33. ("Although varied in their approaches, questions and results, one consistent finding across these studies is that in-movers to gentrifying neighborhoods are wealthier, whiter and of higher educational attainment and out-movers are more likely to be renters, poorer and people of color. The research also consistently shows that rent appreciation predicts displacement.")

25. See, e.g., Andres Duany, "Three Cheers for 'Gentrification,'" *American Enterprise* 12, no. 3 (2001): 36.

26. Alan Mallach, *The Divided City: Poverty and Prosperity in Urban America* (Washington, D.C.: Island Press, 2018), 99.

27. Neil Smith, "Toward a Theory of Gentrification: A Back to the City Movement by Capital, Not People," *Journal of the American Planning Association* 45, no. 4 (1979): 538–548; Richard Florida, "How Urban Core Amenities Drive Gentrification and Increase Inequality," *CityLab*, Dec. 13, 2018, https://www.citylab.com/equity/2018/12/luxury-amenities-gentrification-income-inequality/577924.

28. Kriston Capps, "Study: No Link between Gentrification and Displacement in NYC," *CityLab,* July 31, 2019, https://www.citylab.com/equity/2019/07/gentrification-displacement-link-children-nyc-medicaid-data/594250.

29. Justin Davidson, "New Studies Say Gentrification Doesn't Really Force Out Low-Income Residents," *New York Magazine,* Aug. 5, 2019, http://nymag.com/intelligencer/2019/08/study-gentrification-doesnt-force-out-low-income-residents.html.

30. Rowland Atkinson, "Losing One's Place: Narratives of Neighbourhood Change, Market Injustice and Symbolic Displacement," *Housing, Theory and Society* 32, no. 4 (2015): 373–388.

31. See, e.g., James C. Fraser, Robert J. Chaskin, and Joshua Theodore Bazuin, "Making Mixed-Income Neighborhoods Work for Low-Income Households," *Cityscape* 15, no. 2 (2013): 90 ("Multiple case studies find that residents tend to interact with their neighbors based on perceived characteristics in common"); *Mixed-Income Community Dynamics: Five Insights from Ethnography, U.S. Department of Housing*

and Urban Development, Spring 2013, https://www.huduser.gov/portal/periodicals /em/spring13/highlight2.html.

32. Ana Alvarez, "A New Challenge," *College Hill Independent,* http://www .theindy.org/a/406; Roberta Smith, "Looking for Graphic Lightning from Fort Thunder," *New York Times,* Dec. 6, 2006, https://www.nytimes.com/2006/12/16/arts /design/16wund.html?ex=1168405200&en=a2a23b8ba3abbe0c&ei=5070=.

33. City of Providence Planning and Development, "Woonasquatucket Vision Plan 2018," 49, http://www.providenceri.gov/wp-content/uploads/2017/08/Woonas quatucket-Vision-Plan_Full.pdf.

34. Providence Department of Planning and Development, "Providence Downtown and Knowledge District Plan," Dec. 21, 2012, https://www.providenceri.gov /wp-content/uploads/2017/05/Planning-13.01.23_PD_KD_Plan_100DPI_20121221 .pdf; Elizabeth Abbott, "Providence Puts Focus on Making a Home for Knowledge," *New York Times,* Dec. 13, 2011.

35. Eugenie Birch, "Anchor Institutions in the Northeast Megaregion: An Important but Not Fully Realized Resource," in *Revitalizing American Cities,* ed. Susan Wachter and Kimberly Zeuli (Philadelphia: University of Pennsylvania Press, 2013); Abbott, "Providence Puts Focus on Knowledge."

36. Gail Ciampa, "Gotham Greens Opens Its Massive Providence Greenhouse on Thursday. Here's a Look Inside," *Providence Daily Journal,* Dec. 4, 2019; Grace Kelly, "Greens Manufacturer Begins Operations in Olneyville," *ecoRInews,* Dec. 6, 2019.

37. Jason Schwartz, "End Game: Curt Schilling and the Destruction of 38 Studios," *Boston Magazine,* July 23, 2012, https://www.bostonmagazine.com/news /2012/07/23/38-studios-end-game/; Ben Gilbert, "Former Baseball Star Curt Schilling Just Settled a Lawsuit over His Failed Video Game Studio for $2.5 Million," *Business Insider,* Sept. 20, 2016, https://www.businessinsider.com/curt-schilling-25 -million-settlement-38-studios-2016-9.

38. Casey Ross, "Office Rents Soaring in City's Innovation District," *Boston Globe,* Jan. 10, 2014, https://www.bostonglobe.com/business/2014/01/10/rents-soaring-city -innovation-district/nqeKNcRiLJiyjKEEGog8GP/story.html; see also Meghan Rich and William Tsitsos, "Avoiding the 'SoHo Effect' in Baltimore: Neighborhood Revitalization and Arts and Entertainment Districts," *International Journal of Urban and Regional Research,* 2016, 40 (the creation of an arts and entertainment district in Baltimore was most likely to displace the artists for whom the district was created).

39. Raj Chetty, Nathaniel Hendren, Maggie R. Jones, and Sonya R. Porter, *Race and Economic Opportunity in the United States: An Intergenerational Perspective,* Opportunity Insights, NBER Working Paper No. 24441, March 2018, https://opportu

nityinsights.org/paper/race/?utm_source=newsletter&utm_medium=email&utm
_campaign=newsletter_axioscities&stream=cities.

40. See, e.g., Julie Wagner, "Innovation Districts and Their Dilemmas with Place," Brookings Institution, Feb. 21, 2019, https://www.brookings.edu/blog/the -avenue/2019/02/21/innovation-districts-and-their-dilemmas-with-place/; John Summer, "The People's Republic of Zuckerstan" *The Baffler*, no. 24 (Jan. 2014), https:// thebaffler.com/salvos/the-peoples-republic-of-zuckerstan; Kyle Chayka, "Does Innovation Always Lead to Gentrification?" *Pacific Standard*, June 18, 2014, updated June 14, 2017, https://psmag.com/economics/silicon-valley-disruption-does-innovation -always-lead-to-gentrification-83720.

41. See, e.g., Martin Gilens, *Affluence and Influence: Economic Inequality and Political Power in America* (Princeton, N.J.: Princeton University Press, 2012); Martin Gilens and Benjamin I. Page, "Testing Theories of American Politics: Elites, Interest Groups, and Average Citizens," *Perspectives on Politics* 12, no. 3 (2016): 564–581.

42. For an arresting account of widespread displacement and gentrification in San Francisco, although civic and political leaders throughout the city and region recognized the risk and wanted to avoid it, see Randy Shaw, *Generation Priced Out: Who Gets to Live in the New Urban America* (Berkeley: University of California Press, 2018).

43. Chris Bergenheim, "PBN: Raimondo Nominates 25 R.I. Investment 'Opportunity Zones' to U.S. Treasury Department," HousingWorks RI, Apr. 23, 2018, https://www.housingworksri.org/About/News-and-Events/All-News-Events/Artic leId/230/pbn-raimondo-nominates-25-ri-investment-opportunity-zones-to-us-trea sury-department.

44. "HUD Requests Public Input on How to Maximize Opportunity-Zone Impacts," National Low Income Housing Coalition, Apr. 22, 2019, https://nlihc.org/re source/hud-requests-public-input-how-maximize-opportunity-zone-impacts.

45. Jesse Drucker and Eric Lipton, "How a Trump Tax Break to Help Poor Communities Became a Windfall for the Rich," *New York Times*, Aug. 31, 2019, https:// www.nytimes.com/2019/03/31/business/tax-opportunity-zones.html?action=click &module=Top%20Stories&pgtype=Homepage.

46. Adam Looney, "Will Opportunity Zones Help Distressed Residents or Be a Tax Cut for Gentrification?" Brookings Institution, 2018, https://www.brookings .edu/blog/up-front/2018/02/26/will-opportunity-zones-help-distressed-residents-or -be-a-tax-cut-for-gentrification ("Indeed, the highest returns to investors, and thus the largest tax subsidies will flow to those investing in the fastest gentrifying areas"); see also Hilary Gelfond and Adam Looney, "Learning from Opportunity Zones:

How to Improve Place-Based Policies," Brookings Institution, Oct. 2018, https://www.brookings.edu/wp-content/uploads/2018/10/Looney_Opportunity-Zones_final.pdf.

47. Eric Lipton and Jesse Drucker, "Lawmakers Increase Criticism of 'Opportunity Zone' Tax Break," *New York Times,* Nov. 6, 2018, https://www.nytimes.com/2019/11/06/business/opportunity-zones-congress-criticism.html?action=click&module=Top%20Stories&pgtype=Homepage. If a municipality wants to induce investors to invest in a way that will help the people in these neighborhoods—rather than merely maximize their profit—practically the only option it has is to offer *even more* tax incentives for wealthy developers. That is, the city or state can pile local incentives on top of the federal incentives. Providence, like other cities, has followed this path. It makes widespread use of tax stabilization agreements, which cap property taxes on new developments and guarantee they will not rise for specified periods, which vary with the value of the development. The more valuable the development, the longer the cap remains in place, thus incentivizing developers to undertake larger projects. In exchange for this tax break, developers must agree to comply with the city's rules regarding equal employment, support for minority and women's businesses, and preferential hiring for city residents. But none of this is targeted to Olneyville alone.

48. Lisa Adams, "Erie May Become Model for Federal Opportunity Zones," *Erie News Now,* June 11, 2019, https://www.erienewsnow.com/story/40628936/erie-may-become-model-for-federal-opportunity-zones.

49. "Site Selection: Opportunity Zones," Rhode Island Commerce, https://commerceri.com/site-selection/opportunity-zones.

50. As of mid-2019, the city was in negotiations to sell the complex instead to a group that wanted to create a mixed-use, mixed-income development that would fit with the city's vision for the neighborhood.

51. William Fulton, "Opportunity Zones: Gentrification on Steroids?" Kinder Institute for Urban Research, Feb. 20, 2019, https://kinder.rice.edu/urbanedge/2019/02/20/opportunity-zones-gentrification-steroids.

52. Balazs Szekely, "Downtown LA's 90014 Heads the List of Fastest-Gentrifying ZIPs since the Turn of the Millennium," *RentCafe Blog,* Feb. 26, 2018, https://www.rentcafe.com/blog/rental-market/real-estate-news/top-20-gentrified-zip-codes.

53. Katherine Shaver, "D.C. Has the Highest 'Intensity' of Gentrification of Any U.S. City, Study Says," *Washington Post,* Mar. 19, 2019, https://www.washingtonpost.com/transportation/2019/03/19/study-dc-has-had-highest-intensity-gentrification-any-us-city.

54. These figures are based on real estate sales data in Olneyville provided to me by the city of Providence. Unfortunately, the data set does not distinguish between residential and commercial property sales.

Chapter Ten. Trust the Neighborhood

1. "Mission and Vision," Finger Lakes Land Trust, https://www.fllt.org/about /mission.

2. "About the Finger Lakes Land Trust," Finger Lakes Land Trust, https://www .fllt.org/about.

3. In the classic model, the board of a community land trust consists of an equal number of representatives from three different groups: occupants of CLT housing; residents of the neighborhood where the CLT operates; and outside representatives whose interests align with those of the CLT, including funders, public officials, and nonprofit service providers. For a primer on community land trusts, see, e.g., John Emmeus Davis, "Origins and Evolution of the Community Land Trust in the United States," in *The Community Land Trust Reader* (Cambridge, Mass.: Lincoln Institute of Land Policy, 2010); John Emmeus Davis, "Common Ground: Community-Owned Land as a Platform for Equitable and Sustainable Development," *University of San Francisco Law Review* 51, no. 1 (2014); Oksana Mironova, "How Community Land Trusts Can Help Address the Affordable Housing Crisis," *Jacobin,* July 6, 2019, https://www.jacobinmag.com/2019/07/community-land-trusts-affordable-housing; James DeFilippis et al., "On the Transformative Potential of Community Land Trusts in the United States," *Antipode* 51, no. 3 (2019).

4. "What Is a Community Land Trust?" Center for Community Land Trust Innovation, accessed February 7, 2020, http://www.cltroots.org/what-is-a-clt.

5. John Emmeus Davis, "Dudley Neighbors Inc.," Center for Community Land Trust Innovation, 1988, http://cltroots.org/profiles/dudley-neighbors-inc; James N. Weinstein, Amy Geller, Yamrot Negussie, and Alina Baciu, *Communities in Action: Pathways to Health Equity* (Washington, D.C.: National Academies Press, 2017), 236–239.

6. Weinstein et al., *Communities in Action,* 239–242; Jake Blumgart, "Affordable Housing's Forever Solution," *Next City,* Aug. 10, 2015, https://nextcity.org/features /view/affordable-housings-forever-solution.

7. James DeFilippis et al., "On the Transformative Potential of Community Land Trusts in the United States," *Antipode* 51, no. 3 (2019).

8. Perhaps the best model for such an endowment is the Harlem Children's

Zone (HCZ), which takes a comprehensive approach to the well-being of children and their families within a portion of central Harlem. Largely because of the generous support of a single donor, HCZ dwarfs most community nonprofits. Its annual revenue in 2017 exceeded $150 million, and its endowment approached half a billion dollars. It has used this funding to create and operate more than twenty neighborhood and school programs, including early childhood nutrition and education, K–12 schooling, after-school programs, college preparatory classes, parenting programs, and health education. HCZ has steadily expanded since its founding, from a single block in the 1990s to a ninety-seven-block area today, serving more than twelve thousand children. HCZ, however, is not a land trust and does not decommodify land, leaving Harlem prey to gentrification and displacement.

9. In thinking about the sort of power a neighborhood trust can wield, we can distinguish between what Pierce and Williams call acquisitive resistance and subversion. "Acquisitive resistance involves strategies for gaining power within a relationship with oppressive hegemonies, where subversion involves strategies for dispersing or disrupting relationships that are power-accumulative." Forming a union is an example of the former, since the employees seek to gain power but remain within a preexisting relationship with their employer. By contrast, if the employees leave the business to form their own cooperative, they are exercising subversive power, since they disrupt the power of the original shop and replace old relationships with new ones. Forming a neighborhood trust is an example of subversive power, since it creates a new institution and source of power by disrupting old relationships and replacing them with new ones built around a more egalitarian vision. Joseph Pierce and Olivia R. Williams, "Against Power? Distinguishing between Acquisitive Resistance and Subversion," *Geografiska Annaler: Series B, Human Geography* 98, no. 3 (2016): 171–188, 183.

10. Community chests were the forerunners of the modern United Way. The first community chest was founded in Cleveland in 1913, with the purpose of pooling funds from philanthropy and social services in a single endowment for use by the community, especially its most vulnerable residents. My current home of Ithaca had a community chest that was founded in 1921 and still survives a century later as the United Way of Tompkins County. By 1948 more than a thousand community chests existed across the United States, and today we have more than 1,300 local United Way chapters. See "A Brief History of United Way," United Way of Tompkins County, https://www.uwtc.org/brief-history-united-way.

11. See, e.g., John Emmeus Davis, Rick Jacobus, and Maureen Hickey, "Building Better City-CLT Partnerships: A Program Manual for Municipalities and Commu-

nity Land Trusts," Lincoln Institute of Land Policy, 2008, https://www.lincolninst
.edu/sites/default/files/pubfiles/1401_717_djh_yesim_final.pdf.

12. See Citi Community Development / Grounded Solutions Network Community Land Trust Accelerator, n.d., https://groundedsolutions.org/clt-accelerator
-program.

13. The failure to achieve this alignment is the bane of most funding mechanisms, including tax increment financing. Recent research shows that TIF revenue is often used to provide subsidies for developments in areas that were already moderately successful and "has done little to stimulate growth in the most depressed areas." David Merriman, "Improving Tax Increment Financing for Economic Development," Lincoln Institute of Land Policy, 2018, 57.

14. Sanford F. Schram, *The Return of Ordinary Capitalism: Neoliberalism, Precarity, Occupy* (New York: Oxford University Press, 2015), 31.

15. See Tara J. Melish, "Maximum Feasible Participation of the Poor: New Governance, New Accountability, and a 21st Century War on the Sources of Poverty," *Yale Human Rights and Development Journal* 13, no. 1 (2010).

16. Mehrsa Baradaran argues persuasively that reforms that do no more than bring capitalism to distressed neighborhoods, without reckoning with the long and continuing history of discrimination in access to credit markets, will never change the status quo. See Mehrsa Baradaran, "Jim Crow Credit," *UC Irvine Law Review* 9, no. 4 (2019): 887, https://scholarship.law.uci.edu/ucilr/vol9/iss4/4. Among the reforms Baradaran calls for is the creation of shared-equity mortgages to give the poor greater access to homeownership (947). These mortgages divide ownership between an investor and a homeowner but do not place limits on the homeowner's profit should she decide to sell on the open market. As a result, Baradaran's plan would not guarantee long-term affordability.

17. See, e.g., Mehrsa Baradaran, *The Color of Money: Black Banks and the Racial Wealth Gap* (Cambridge, Mass.: Harvard University Press, 2019); Justin Gomer, "Housing and the Racial Wealth Gap: A Historical Overview," KCET.org, Sept. 4, 2018, https://www.kcet.org/shows/city-rising/housing-and-the-racial-wealth-gap-a
-historical-overview.

18. Baradaran, *The Color of Money*; Richard Rothstein, *The Color of Law: A Forgotten History of How Our Government Segregated America* (New York: Liveright, 2017).

19. Scholars refer to this shift as a "transformation of subjectivities . . . when people imagine themselves as actors not primarily driven by economic motives and rationalities. In the case of housing, it may also occur when they imagine their house-

holds as part of a broader polity or community, one framed around housing and locale." James DeFilippis et al., "On the Transformative Potential of Community Land Trusts in the United States," *Antipode* 51, no. 3 (2019).

20. On the great diversity within Latino communities, see, e.g., Ronald Schmidt Sr., Edwina Barvosa-Carter, and Rodolfo D. Torres, "Latino/a Identities: Social Diversity and U.S. Politics," *PS: Political Science and Politics* 33, no. 3 (2000): 563–567. Many scholars have found comparable diversity within Black communities. See, e.g., Mary Patillo, *Black on the Block: The Politics of Race and Class in the City* (Chicago: University of Chicago Press, 2007).

21. Alan Mallach, *The Empty House Next Door: Understanding and Reducing Vacancy and Hypervacancy in the United States,* Lincoln Institute of Land Policy, 2018, 4, 7, https://www.lincolninst.edu/sites/default/files/pubfiles/empty-house-next-door -full.pdf.

22. Joseph Margulies, "Communities Need Neighborhood Trusts," *Stanford Social Innovation Review,* Spring 2019, https://ssir.org/articles/entry/communities_need _neighborhood_trusts#.

23. Jennifer Percy, "Trapped by the 'Walmart of Heroin,'" *New York Times Magazine,* Oct. 10, 2018, https://www.nytimes.com/2018/10/10/magazine/kensington-hero in-opioid-philadelphia.html.

24. Percy, "Trapped by the 'Walmart of Heroin.'"

25. See, e.g., Bruce Katz, "The New Street Fight," *Philadelphia Citizen,* Sept. 20, 2019, https://thephiladelphiacitizen.org/shift-capital-kensington/; Oscar Perry Abello, "New Real Estate Trust Wants Troubled Philly Community to Shape Its Own Economic Development," NextCity.org, Nov. 19, 2019, https://nextcity.org /daily/entry/real-estate-trust-wants-troubled-philly-community-shape-own-devel opment; Amy Cortese, "Neighborhood Investment Company Helps Los Angeles Residents Buy Back the Block," Impactalpha.com, Dec. 4, 2019, https://impactalpha .com/neighborhood-investment-company-helps-los-angeles-residents-buy-back -the-block.

Epilogue

1. Employment Situation Summary, Bureau of Labor Statistics, May 8, 2020, tables A-2, A-3, https://www.bls.gov/news.release/empsit.nro.htm.

2. Tracy Jan and Scott Clement, "Hispanics Are Almost Twice as Likely as Whites to Have Lost Their Jobs amid Pandemic, Poll Finds," *Washington Post,* May 6, 2020, https://www.washingtonpost.com/business/2020/05/06/layoffs-race-poll-cor

onavirus/; "Labor Force Statistics from the Current Population Survey," U.S. Bureau of Labor Statistics, 2019, https://www.bls.gov/cps/cpsaat18.htm.

3. Board of Governors of the Federal Reserve System, "Report on the Economic Well-Being of U.S. Households in 2018," May 2019, 2, https://www.federalreserve.gov/publications/files/2018-report-economic-well-being-us-households-201905.pdf.

4. Jeanna Smialek, "Poor Americans Hit Hardest by Job Losses amid Lockdowns, Fed Says," *New York Times,* May 14, 2020, https://www.nytimes.com/2020/05/14/business/economy/coronavirus-jobless-unemployment.html.

5. Sam Van Pykeren, "These Photos Show the Staggering Food Bank Lines across America," *Mother Jones,* Apr. 13, 2020, https://www.motherjones.com/food/2020/04/these-photos-show-the-staggering-food-bank-lines-across-america.

6. Jennifer Valentino-DeVries, Denise Lu, and Gabriel J. X. Dance, "Location Data Says It All: Staying at Home during Coronavirus Is a Luxury," *New York Times,* Apr. 3, 2020, https://www.nytimes.com/interactive/2020/04/03/us/coronavirus-stay-home-rich-poor.html; J. C. Pan, "Essential Workers, Disposable Jobs," *New Republic,* Apr. 17, 2020, https://newrepublic.com/article/157334/essential-jobs-disposable-grocery-workers-coronavirus; Catherine Powell, "Color of COVID: The Racial Justice Paradox of Our Stay-at-Home Economy," CNN.com, Apr. 18, 2020, https://www.cnn.com/2020/04/10/opinions/covid-19-people-of-color-labor-market-disparities-powell/index.html; Elise Gould and Heidi Shierholz, "Not Everybody Can Work from Home," Economic Policy Institute, Mar. 19, 2020, https://www.epi.org/blog/black-and-hispanic-workers-are-much-less-likely-to-be-able-to-work-from-home.

7. Employee Benefits in the United States, Bureau of Labor Statistics, table 2, "Medical Care Benefits: Access, Participation, and Take-Up Rates"; table 6, "Selected Paid Leave Benefits," Mar. 2019, https://www.bls.gov/news.release/pdf/ebs2.pdf.

8. Dhruv Khullar and Dave A. Choksi, "Health, Income, and Poverty: Where We Are and What Could Help," *Health Affairs,* Oct. 4, 2018, https://www.healthaffairs.org/do/10.1377/hpb20180817.901935/full.

9. Miriam Jordan and Richard A. Oppel Jr., "For Latinos and COVID-19, Doctors Are Seeing an 'Alarming' Disparity," *New York Times,* May 8, 2020, https://www.nytimes.com/2020/05/07/us/coronavirus-latinos-disparity.html.

10. John Eligon, Audra D. S. Burch, Dionne Searcey, and Richard A. Oppel Jr., "Black Americans Face Alarming Rates of Coronavirus Infection in Some States," *New York Times,* Apr. 14, 2020, https://www.nytimes.com/2020/04/07/us/coronavirus-race.html; Reis Thebault, Andrew Ba Tran, and Vanessa Williams, "The Coronavirus Is Infecting and Killing Black Americans at an Alarmingly High Rate," *Washington*

Post, Apr. 7, 2020, https://www.washingtonpost.com/nation/2020/04/07/corona
virus-is-infecting-killing-black-americans-an-alarmingly-high-rate-post-analysis
-shows/?arc404=true.

11. Rhode Island Department of Health, COVID-19 Rhode Island Data, https://
docs.google.com/spreadsheets/d/1n-zMS9Al94CPj_Tc3K7Adin-tN9x1RSjjx2UzJ4
SV7Q/htmlview#.

12. Rhode Island Department of Health, COVID-19 Rhode Island data; New
York City Health Department, COVID-19 data, https://www1.nyc.gov/site/doh/co
vid/covid-19-data.page.

13. Trust for America's Health, "The Impact of Chronic Underfunding on
America's Public Health System: Trends, Risks, and Recommendations, 2020," Apr.
2020, https://www.tfah.org/wp-content/uploads/2020/04/TFAH2020PublicHealth
Funding.pdf.

14. Trust for America's Health, "Impact of Chronic Underfunding, 2020."

15. Trust for America's Health, "The Impact of Chronic Underfunding on
America's Public Health System: Trends, Risks, and Recommendations, 2019," Apr.
2019, https://www.tfah.org/wp-content/uploads/2020/03/TFAH_2019_PublicHea
lthFunding_07.pdf; Trust for America's Health, "A Funding Crisis for Public Health
and Safety: State by State Public Health Funding and Key Health Facts, 2018," Mar.
2018, https://www.tfah.org/report-details/a-funding-crisis-for-public-health-and-sa
fety-state-by-state-and-federal-public-health-funding-facts-and-recommendations.

16. "Coronavirus (COVID-19): Revised State Revenue Projections," National
Conference of State Legislatures, n.d., https://www.ncsl.org/research/fiscal-policy
/coronavirus-covid-19-state-budget-updates-and-revenue-projections637208306.as
px#FY%202020.

17. Dean Mosiman, "Madison Expects $30M Budget Shortfall, Resulting Cut-
backs," *Wisconsin State Journal,* May 6, 2020; Aaron Besecker, "Buffalo Might Run
Out of Money If Feds Don't Provide Aid," *Buffalo News,* Apr. 29, 2020; "The COV-
ID-19 Effect on San Francisco's Budget," *sf.citi,* Apr. 16, 2020, https://sfciti.org/news
/the-covid-19-effect-on-san-franciscos-budget/; Tony Romm, "Mass Layoffs Begin
in Cities and States amid Coronavirus Fallout, Threatening Education, Sanitation,
Health and Safety," *Washington Post,* Apr. 29, 2020, https://www.washingtonpost
.com/business/2020/04/29/cities-states-layoffs-furloughs-coronavirus.

18. John MacIntosh, "COVID-19 Could Mean Extinction for Many Charities,"
CNN.com, Mar. 20, 2020, https://www.cnn.com/2020/03/20/opinions/coronavirus
-extinction-level-event-charities/index.html; "The Voice of Charities Facing COV-

ID-19 Worldwide," *CAF America,* Apr. 6, 2020, https://www.cafamerica.org/wp-con tent/uploads/CV19_Report_CAF-America.pdf; "'We Had a Shoestring Budget in Good Times': COVID-19's Devastating Impact on the Nonprofit Sector," *Catchafire Blog,* Mar. 23, 2020, https://catchafireblog.org/we-had-a-shoestring-budget-in-good -times-covid-19-s-devastating-impact-on-the-nonprofit-sector-434df2d5f78b.

Acknowledgments

Each of my books tries to answer a single question. Fortunately, it's not the same question, since that would bore me, infuriate my wife, and annoy my publisher. But I know I'm not ready to write until I can distill my research into one clear question. In the book you are reading, the question is this: Can the low-income residents of a distressed neighborhood make their home safe, vibrant, and healthy without setting the stage for gentrification and displacement? Getting me to the point where I finally understand and can state the question at the core of my research takes a lot of work by a lot of people, and it is my great pleasure to thank some of them here.

Foremost, this book is about the interaction between people and place, and my greatest debt is to the men and women who live and work in Olneyville. When I started my research, I heard time and again that Olneyville was a special neighborhood. The more time I spent there, the more I knew it was true, and I count myself lucky for having found it. I hope I have done it justice in these pages. More than that, I hope my work can help preserve it.

I interviewed a great many people for this book, some at considerable length. I would like to thank Adriana Abizadeh, Valerie Almeida-Monroe, Kyle Bennett, Mercedes "Betty" Bernal, Jen Chapman, Channavy Chhay, Brian Chippendale, David Cicilline, Marlon Cifuentes, Hugh Clements, Eliza Cohen, Carla Cuellar, Colleen Daley Ndoye, Roshni Darna, Thomas Deller, Ian Donnis, Jorge

Elorza, Susan Erstling, Barbara Fields, Paul Fitzgerald, Stephanie Fortunato, Julia Gold, Deb Golding, Jennifer Hawkins, Jungil Hong, Nancy Howard, Dilania Inoa, David Kemper, Cate Latz, Alicia Lehrer, Robin Levasseur, Susan Lusi, Xander Marro, Paul Mazarella, Sophie McConnell, Robert McMahon, Yuselly Mendoza, Virginia Morgan, Chris Morrison, Brian Murray, Ana Novais, Casey O'Donnell, Fred Ordoñez, Martin "Big Daddy" Pagan, Francis Parra, Mia Patriarcha, Jennifer Pereira, Walesca Pinto, Jacqueline Reyes, April Ricci, Lily Rivera, Chris Rotondo, Brent Runyon, Allegra Scharff, Andrew Schiff, Kavya Shankar, Tina Shepard, Jesse Sivak, Howie Sneider, Eric Stewart, Eric Stover, Meg Sullivan, Anthony Taylor, Father Raymond Tetrault, Mark Van Noppen, Cecil Vega, and April Wolf.

Some people met with me repeatedly during my research, and their wisdom contributed to nearly every page of this book. I could not have written it without Lisa Aurecchia, Dean Esserman, Richard Fernandes, Iasha Hall, Abelardo Hernandez, Dean Isabella, Brent Kerman, Sabina Matos, Bonnie Nickerson, Delia Rodriguez-Masjoan, Frank Shea, Jane Sherman, and Elmer Stanley. They spent hours trying to help me understand their work and passion for a small neighborhood on the west side of town. A special nod goes to Bonnie Nickerson, director of the Providence Department of Planning and Development, who was endlessly gracious even though she disagrees with my conclusions. She is a true professional and a great asset to the people of Providence.

In the course of my research, I spent the better part of a week working at the Mary Elizabeth Robinson Research Center in Providence, where I was given access to the collections of the Rhode Island Historical Society. The staff there helped me dig out historical treasures about Olneyville that I could not have located elsewhere, and I thank them for their help. I also spent a glorious two days in the

Providence City Archives, tucked away on the top floor of Providence City Hall. There I found things about Olneyville that I was certain no longer existed. I am grateful to the staff for their assistance.

This is my second book with Yale University Press and executive editor Bill Frucht. It is an honor to publish with Yale and a delight to work with Bill. He saw and understood my core question even before I did, and I appreciate his insight. I am also grateful to Bill Henry, who brought his uncanny eye for detail to the grossly underappreciated job of copyediting, and to Mary Pasti, whose alchemy magically transformed a manuscript into a book. She assures me it took a village, and I extend my thanks to her colleagues as well. I am fortunate to be represented by Lisa Adams of the Garamond Agency, an immensely patient literary agent. I just hope she never calculates her fee on an hourly basis.

One of the joys of working at a place like Cornell, my professional home for the last seven years, is the opportunity to learn from wonderful students. Ella Bcublik and Sawyer Smith provided research at an early stage of the project, and I appreciate their help. As the book took shape, I had the good fortune to work closely with Grace Mehler and Hannah Ambinder, remarkable students whose assistance in the last year of writing was invaluable. Both have since graduated and are on their way to brilliant careers in the law. They share an enviable commitment to social justice, and, mark my words, you will hear their names again. And in a truly fortuitous twist, I was lucky to teach Marisa O'Gara at Cornell Law School. Before coming to Cornell, Marisa had been chief of staff for Providence mayor Jorge Elorza and had an insider's perspective on both the actors and the issues facing Olneyville. She read the entire manuscript and provided invaluable feedback.

Cornell is the home of great scholars, and two in particular took a benevolent interest in my research. Nancy Brooks patiently explained the complex economics of U.S. housing policy and graciously agreed to read the manuscript, after which she provided a raft of recommendations, all of which I prudently adopted, and some of which saved me from embarrassing error. But my greatest scholarly debt goes to my friend and colleague Jamila Michener. She is a scholar extraordinaire with a formidable expertise in the politics of poverty. She is also a kind and supportive colleague who encouraged me throughout my research. She too reviewed the entire manuscript and provided page after page of detailed, constructive comment. It is no exaggeration to say that this book would not be what it is were it not for her guidance.

As everyone who has ever been foolish enough to write a book can attest, you cannot do it without pals to prop you up. I am particularly fortunate on that score because one of my dearest friends also happens to be among the best nonfiction writers alive. Alex Kotlowitz never tired of listening to me talk about Olneyville, or if he did, he had the decency to pretend otherwise. I will never write like he does (when I tell him that, he says, "But Joe, I'll never write like you do"; I get the feeling he says that with considerable relief), but if my writing is good, it is because Alex made it better. His encouragement came at all the right moments, and I am much in his debt.

Finally there is Sandra. She is the best thing that ever happened to me. I hope she knows I needn't say more.

Index